T0136081

About Island Press

Since 1984, the nonprofit Island Press has been stimulating, shaping, and communicating the ideas that are essential for solving environmental problems worldwide. With more than 800 titles in print and some 40 new releases each year, we are the nation's leading publisher on environmental issues. We identify innovative thinkers and emerging trends in the environmental field. We work with world-renowned experts and authors to develop cross-disciplinary solutions to environmental challenges.

Island Press designs and implements coordinated book publication campaigns in order to communicate our critical messages in print, in person, and online using the latest technologies, programs, and the media. Our goal: to reach targeted audiences—scientists, policymakers, environmental advocates, the media, and concerned citizens—who can and will take action to protect the plants and animals that enrich our world, the ecosystems we need to survive, the water we drink, and the air we breathe.

Island Press gratefully acknowledges the support of its work by the Agua Fund, Inc., The Margaret A. Cargill Foundation, Betsy and Jesse Fink Foundation, The William and Flora Hewlett Foundation, The Kresge Foundation, The Forrest and Frances Lattner Foundation, The Andrew W. Mellon Foundation, The Curtis and Edith Munson Foundation, The Overbrook Foundation, The David and Lucile Packard Foundation, The Summit Foundation, Trust for Architectural Easements, The Winslow Foundation, and other generous donors.

The opinions expressed in this book are those of the author(s) and do not necessarily reflect the views of our donors.

SHIFTING BASELINES

Shifting Baselines

THE PAST AND THE FUTURE OF OCEAN FISHERIES

EDITED BY

Jeremy B. C. Jackson, Karen E. Alexander,
and Enric Sala

Washington | Covelo | London

Figures designed by Sherry Palmer

Library of Congress Cataloging-in-Publication Data

Shifting baselines : the past and the future of ocean fisheries / edited by Jeremy B.C. Jackson Karen Alexander, and Enric Sala.
p. cm.
Includes bibliographical references and index.
ISBN-13: 978-1-61091-000-2 (hardback)
ISBN-10: 1-61091-000-1 (cloth)
ISBN-13: 978-1-61091-001-9 (paper)
1. Fisheries—History. 2. Fishery management. I. Jackson, Jeremy B. C., 1942–
II. Alexander, Karen, 1951 III. Sala, Enric.
SH211.S45 2011
338.3′727—dc22
2011005032

Printed on recycled, acid-free paper

Manufactured in the United States of America
10 9 8 7 6 5 4 3 2 1

Keywords: fisheries management, marine ecosystems, biodiversity, historical ecology, maximum sustainable yield, fishing down the food web, anchovy, sardine, cod

We dedicate this book to our friend and colleague Daniel Pauly for his fundamental insight in how problems of shifting baselines shape and distort our lives.

CONTENTS

Introduction: The Importance of Shifting Baselines

JEREMY B. C. JACKSON AND KAREN E. ALEXANDER

Exploding out of Africa just seventy thousand years ago, human beings colonized every continent but Antarctica before the end of the last ice age. In the process, we drove three-quarters of the land animals larger than 100 pounds extinct—a truly remarkable achievement for fewer than five million people armed with sticks and rocks. The oceans were still largely safe, but heaps of bones and shells along the shoreline augured what was soon to come.

Then about ten thousand years ago, people began to settle down and invent agriculture and trade, towns, cities, and bureaucracies. Armies soon followed. The familiar pattern of conquest, expansion, environmental destruction, and the rise and fall of empires was firmly established a few thousand years later. Fishing also intensified, and the numbers of shellfish, fish, sea turtles, sea cows, and seals began to drop worldwide. Persian stone friezes in the Louvre vividly depict abundant sea turtles, marine mammals, and fish at the time of Sargon the Great, around 700 BCE, but by the time of the Romans most of these animals were growing scarce. Romans cast their nets and built factories all the way from the northern Black Sea to Britain to meet an ever-growing demand for fish and fish products. Similar stories abound from the Americas, where middens tell us that people shifted to tinier and tinier prey for subsistence, or they targeted fish that were harder to catch or farther away. The human population was somewhere between two and three hundred million.

Fast-forward to the great maritime empires of the fifteenth to nineteenth centuries, when the oceans became a vast new fishing ground and a superhighway to move people and commodities. Bigger and better ships took to the open oceans to catch herring, cod, and great whales en masse. One of the first casualties was the Atlantic gray whale, hunted to extinction by the eighteenth century. Herring and cod were said to be inexhaustible. Yet by the end of the nineteenth century, ships were fishing farther and farther out to sea with increasingly sophisticated gear, but catches of even the mighty cod showed evidence of steep decline. Similar stories abound for oysters, shad, and alewives from New England to the Chesapeake Bay, and sea turtles and monk seals from the Caribbean. Chemical pollution and invasions of nonnative species also increased, and entire estuaries and coastal ecosystems were devastated by 1900. Meanwhile, human population increased to about 1.5 billion by the end of the nineteenth century.

But the real damage had only just begun. What had previously been a series of local or regional problems were rapidly becoming global. The driving force was our relentless quest for progress and the necessity for growing economies to feed, govern, and placate increasing billions of people. The engine ran on cheap energy from a seemingly endless supply of fossil fuels. Despite the carnage of two great world wars, humans increased to 2.5 billion by 1950.

Since then, the oceans as we knew them have begun to die. Most of the largest fish are gone and, according to the latest conservative estimate, more than 80 percent of the world's major industrial fisheries have crashed or are over- or fully exploited. Sports fishers pay more and more money to catch fewer and smaller fish. Apex predators like tuna, salmon, and swordfish—and the people who eat them—are increasingly full of mercury, dioxins, and PCBs. Gigantic amounts of plastics are trapped in ocean gyres, and dead zones of hypoxic waters have increased from a few dozen in the 1950s to more than four hundred today. Reef corals are dying en masse from outbreaks of bleaching and disease fueled by rising temperatures, and the acidification of surface waters due to increased carbon dioxide threatens virtually all sea life with calcareous skeletons, including corals, shellfish, and plankton.

As a result, entire ecosystems are in danger of extinction. Coral reefs, estuaries, and coastal seas are "critically endangered" globally. Vast fishing grounds of the continental shelves and seamounts are "endangered" and the open ocean pelagic realm is "threatened." Meanwhile, humans have nearly tripled to more than 6.5 billion, and we have increased our consumption of the total renewable resources the earth can provide annually—the so-called global ecological footprint—from less than 50 percent in 1950 to 150 percent today. We are living off of our ecological credit cards and the interest rates are going up.

All of this is well known to historians and ecologists, but that knowledge hasn't changed natural resource policy. We are blithely managing marine extinctions by ignoring past failures. This dangerous lack of historical perspective was the stimulus for Daniel Pauly's brilliant 1995 essay titled "Anecdotes and the Shifting Baselines Syndrome in Fisheries," which marked a fundamental turning point in conservation biology and fisheries science. Pauly's basic point was that we have lost sight of nature because we ignore historical change and accept the present as natural. Understanding the perilous state of fisheries would require historical perspective to determine the true magnitude of decline and the challenges for sustainability in the future.

Shifting baselines is a truly fundamental and revolutionary idea, but the revolution has not yet happened because the challenges are enormous. There are at least three impediments to change. First, it is not enough to measure only what we see today because some of the most important changes happened before scientists began to measure them. Consequently, it is essential to adopt a truly interdisciplinary approach, using a wide variety of data to estimate past changes and understanding those changes in a social and historical as well as scientific context. Second, shifting baselines challenges long-established goals for management that were based on simplistic concepts such as maximum sustainable yield (MSY). It illustrates how past practices have destroyed healthy ecosystem structure and function, lessons that must now be incorporated into fisheries management. Third, shifting baselines makes us uncomfortable because it places all of us squarely within nature and holds us accountable for both past destruction and shaping the future.

This book is a first joint attempt by scientists and historians to explore the significance of the shifting baselines paradigm. What does it mean for the future of fisheries and the ways in which we perceive our ever more unnatural oceans? It is neither a comprehensive synthesis of scientific papers about the collapse of fisheries nor a fisheries history. This information has been ably covered elsewhere. Rather, it shows how new perspectives on the past can alter our understanding of oceans today and change the future for the better.

To achieve this, we need to establish a minimum set of parameters required for basic understanding that excludes superfluous detail but is amenable to appropriate changes in spatial and temporal scale. Such an approach is essential to answer three practical and important questions. First, how much, and in what ways, have marine ecosystems changed because of human impacts, as opposed to natural changes? Second, what were the trajectories, scale, and tempo of change, and how can we distinguish between cause and effect? Third, how can we use insights from historical ecology to ameliorate the degradation of marine resources and biodiversity? We begin here with two well-studied fisheries, but the methods apply across the full range of human impacts on watersheds, coastal regions, and open oceans. We firmly believe there is no hope of success without historical perspective.

In 1968, Richard Levins observed that "it is not possible to maximize simultaneously generality, realism, and precision." Following his maxim, historical ecology commonly sacrifices precision for generality and realism. General principles emerge from case studies that describe and predict the long-term consequences of overfishing or habitat destruction. This has been accomplished for degraded coral reefs and for estuaries and coastal seas around the world. Essentially, the process is the same: general patterns of degradation are repeated over and over again. First the big animals are wiped out, then the smaller ones. Large herbivores, usually slower, safer targets, generally disappear before large carnivores, usually faster, wilier, more dangerous prey. Habitat structure is imperiled once large animals have been removed. When you have seen one degraded coral reef or estuary, you have seen them all, not in fine detail, but in terms of process. There is an important message in this sameness.

In contrast, most ecologists, fisheries biologists, policymakers, and fishers today focus on quantitative estimates of population size rather than on functional processes. Conventional scientific wisdom tells us that historical data are rarely precise enough to estimate past populations (although evidence mounts to the contrary), so realism is sacrificed for precision. But such "precisionism" is seriously misguided. It focuses on recent fluctuations of a few percent while ignoring extraordinary losses in the past. We miss the signal by focusing intently on what is all too commonly statistical noise.

Realistically, marine scientists need to know about long-term changes in species abundance and distribution. Which species that were once abundant are now extinct or vastly diminished? How have ranges contracted or concentrations become diffused? How has essential habitat changed? What kinds of organisms have filled vacant ecological niches? How has the topology of food webs changed? Answers to these questions provide the best evidence we can hope for in anticipating the consequences of conservation actions such as stopping fishing entirely, restricting specific gear, setting catch limits, or establishing large marine protected areas.

The story of the Caribbean provides a model for how to answer these questions. Voyages of discovery spearheaded imperial adventures for economic gain and power. Explorers, starting with Columbus, were good observers and practical folk. They had an eye for commodities that they could sell or they would not have been explorers for long. Civil servants followed, such as Fernández de Oviedo, whose *General and Natural History of the Indies*, published in 1534, is an impressively objective executive summary for the king of Spain about the natural resources of the Caribbean. That Oviedo chose to emphasize sea cows, green turtles, and sharks above all other marine creatures, animals that are now ecologically extinct throughout the Caribbean, speaks loudly about their extraordinary abundance, value, and danger in the 1500s. Oviedo cataloged not just these big animals, but also fish, sponges, lobsters, conchs, and sea cucumbers.

Historical data alter the scale of abundance and distribution. It matters very much that in Oviedo's day there were between fifty and a hundred nesting beaches with enormous numbers of Caribbean green turtles because, heretofore, scientists assumed there were fewer than twenty nesting sites. Oviedo's observation implies that there were at least 50–100 million adult turtles in the Caribbean, when nobody had imagined more than 1 or 2 million before. It matters for the management of Caribbean seagrass ecosystems that these millions of green turtles, a lot bigger on average than those today, ate proportionately more seagrass. The future survival of seagrass ecosystems may depend upon restoring much larger populations of these animals than conventional marine science could have predicted. Locating and resurrecting the most viable historical nesting sites will be essential to restoring both the turtle populations *and* the seagrass ecosystems.

This book grew out of a conference in November 2003 at the Scripps Institution of Oceanography—the second of three conferences on *Marine Biodiversity: The Known, Unknown, and Unknowable*. Papers and discussion at the conference focused on varieties of past evidence arranged in long-term data sets to produce scientific results, but notions of certainty and uncertainty divided the assembly, and the same concepts voiced by ecologists and historians often carried radically different meanings. However, two topics dominated the media and were on everyone's mind: the recent global decline of large predatory fish and the collapse of the Newfoundland cod fishery after five hundred years of commercial fishing. Scale matters, and these large-scale events colored the debate.

The debate has evolved significantly since 2003, and this book reflects current events that make the fundamental issues timelier. The chapters are arranged in five parts: the problem statement, two sets of case studies, methods, and discussion. Thematically arranged, each part is framed by an introduction from the editors and displays a narrative tension that reflects real differences in the philosophy, perspective, and academic disciplines of the authors. We structured the book to highlight these conceptual differences as a first step toward resolving them.

Rather than recapitulate findings already published elsewhere, several authors used the book as an opportunity to reflect on the field and their work with candor and introspection. They examined process in science and history and speculated about the significance of collaboration in an uncertain future. The results are often introspective and humane, and points of view converge as often as they differ.

Part I presents three perspectives on the problem of shifting baselines. In Carl Safina's elegy, a single lifetime stands witness to profound local environmental change, and personal memory stands in for the historical past. In contrast, Rashid Sumaila and Daniel Pauly remind us that governments around the world continue to embrace folly rather than sound policies when it comes to fisheries on the brink of collapse. They then slide the baseline from the past into the future to illustrate the peril of economic and environmental presentism, a theme recapitulated in the concluding chapter. Marine biologist turned filmmaker Randy Olson argues that scientists must communicate more effectively with the public if the oceans are to be saved.

The first case study, the anchovy-sardine conundrum in Part II, has engaged marine science since the middle of the twentieth century. The story exhibits the extraordinary complexity and nonstationary dynamics of the Pacific Ocean. As both Alec MacCall and David Field and colleagues point out, the problem turned out to be historical as well as geographical in scale. Fisheries scientists in the 1980s and 1990s linked novel data sets outside their normal purview to show that the fish populations responded to decadal-scale climate cycles, a pattern impossible to detect using traditional sampling surveys and landings records.

In contrast, Part III shows there is no clear case implicating oceanographic factors in the demise of northwest Atlantic cod. The story is one of historical detective work going back several centuries. Jeff Bolster, Karen Alexander, and Bill Leavenworth describe how historians developed a time series of total removals based on a contextual analysis of landings from centuries-old logbooks, which was then analyzed statistically using a fisheries stock assessment model. Nevertheless, in his memoir of Newfoundland before and after the collapse of the codfishery, Daniel Vickers cautions that historical data are always filtered and interpreted through individual experience.

The chapters in Part IV discuss the myriad kinds of data and methodologies employed to assess the degradation of coastal seas and the pristine population sizes of great whales. As reviewed by Heike Lotze and colleagues, proxies for the past are imperfect. Our confidence increases when different proxies and methods agree and falls when they do not. Stephen Palumbi reemphasizes the cautions expressed by Vickers about historical analysis by comparing wildly different estimates of past whale abundance based on molecular population genetics versus others based on historical whaling records. Here we are pushing at the very boundaries of what is known versus what is yet unknown in historical analysis.

The essays in Part V stress the importance of historical perspectives for effective management. First, Andrew Rosenberg, Karen Alexander, and Jamie Cournane discuss fisheries management in New England, an area famous for confrontation. Rosenberg writes from long, personal experience on the front lines of United States policy and government service, while Cournane is beginning a career in fisheries management. Enric Sala and Jeremy Jackson elaborate on their coral reef studies to outline how these different insights can be brought to bear to manage the ocean's future. The epilogue brings us up to date in 2010.

The Alfred P. Sloan Foundation and its related programs in the History of Marine Animal Populations and the Census of Marine Life, the Center for Marine Biodiversity and Conservation (CMBC), and the Scripps Institution of Oceanography sponsored the conference. Sarah Mesnick helped to develop the scientific program, tracked down the participants, and coordinated the participation of graduate students in every aspect of the symposium. She and the logistical genius of Penny Dockry made the symposium possible. We also thank Nancy Knowlton, who made CMBC such a nurturing place to work, Ivan Gayler and the late Alan Jaffe for their kind support, and Jesse Ausubel, for believing that historical perspective is fundamental to facing the environmental challenges of the future.

At Island Press, Todd Baldwin shepherded the manuscript through the review process, and Emily Davis oversaw all aspects of publication. Sharis Simonian took the book through production, and Jaime Jennings handled publicity. Their efforts and the comments of our anonymous reviewers greatly improved the finished product. We are grateful to them all.

PART I

The Problem Defined

In the three introductory essays, Carl Safina, Rashid Sumaila and Daniel Pauly, and Randy Olson frame the fisheries crisis in human terms. Safina poses the epistemological questions. Why is the past important? Is history like memory? Does it provide a necessary context for decision making? If this is true, does it follow that institutions ignorant of their history behave like people with impaired memory, confronting recurring dilemmas as entirely new? Can and should knowledge of the past influence modern marine science and policy, and in what way?

Sumaila and Pauly look to the future and predict that, unless behavior changes, humankind will continue on the "march of folly" of fisheries. Using Barbara Tuchman's famous metaphor, they show how knowledge of the past, ignored in the past, established pernicious fisheries policies that actually worked against the best interests of the majority of people and exhibited a venal indifference to the well-being of future generations.

Olson's approach is utilitarian. His forum is mass media, his audience the digital generation. People must be convinced to modify their behavior if the oceans are to be restored, and the stakes are too high to rely on message alone to convince them. Packaging the message is equally important in the digital media age. Using case studies, he explains which media campaigns worked, which didn't work, and why. Then he outlines how to effectively communicate marine science to the public.

A marine scientist by training, Safina's most recent professional publications have been on the need to add teeth and resolve to fisheries management, and his celebrated books have instilled concern about the ocean's condition in a wide general audience. Here his essay takes a deeply personal approach to shifting baselines. Most academic scientists move about like gypsies and have missed witnessing firsthand the slow, but profound changes taking place almost everywhere. In a memoir of the Long Island shore he has known since childhood, Safina confronted the process in his own lifetime. He reminds us that the importance of place is not only abstract, scientific, and historical but also intimate and tangible. The small scale resonates most clearly with human experience, and the individual is still a fulcrum that can shift the world.

Sumaila and Pauly advance economic theories of resource allocation that advocate fair distribution to future generations and undercut policies that support overfishing worldwide. As an economist, Sumaila has worked on natural resource allocation and policy development all around the world. Pauly has published widely in all areas of fisheries science, but increasingly has focused attention on the role of fisheries in providing food and self-sufficiency to poor and marginalized people, particularly in Africa and Asia. Like Voltaire, he is known for distilling fundamental concepts into a few memorable words. The authors framed their essay around a historian's memorable words and employed citations that marshal an impressive array of scientific and economic papers as evidence. Tuchman would be amused to find her historical concept supported by so many statistical models. Yet the point of the essay is not the past, but the future. We inherited damaged marine ecosystems because we are the heirs of past bad planning. Sumaila and Pauly challenge us to do better for future generations by implementing policies that history and science have shown may be successful.

Now a filmmaker, Olson was once a marine biologist. His first film, *Lobstahs*, was about lobster fishing and fishermen in the Gulf of Maine. Since then he has worked with Jeremy Jackson on the short film *Re-Diagnosing the Oceans* and on the Shifting Baselines Ocean Media Project, using humor to communicate to the public the alarming state of the oceans. His recent films, *Flock of Dodos: The Evolution–Intelligent Design Circus* (2006) and *Sizzle: A Global Warming Comedy* (2008), and his book *Don't Be Such a Scientist* (2009) use edgy humor to criticize scientists and science foundations for failing to effectively communicate with the public on issues of critical importance. His productions don't look like business as usual, and he has drawn scathing rebukes from many in the scientific establishment for his brash, irreverent approach—although never for his science. Olson explains why innovative public communication is vitally important for the future of the oceans and challenges the establishment to get with the program.

Chapter 1

A Shoreline Remembrance

Carl Safina

Orienting Memories

Well over forty years ago—I was about five—my father drove us from Brooklyn to Long Island for a day of picnicking and fishing at a coastal state park. At one point, my mother bravely walked me into the edge of a gull colony. A city girl from Manhattan, she must have been almost as frightened as I, because I remember her squeezing my hand and holding her hat down as the birds—seemingly the size of condors—swooped in with menacing threat calls and the close whoosh of wings. I was terrified. But then suddenly at my feet, there was an amazing bowl of grass and feathers cradling three astonishing, huge speckled eggs. It was my first brush with something wild, and it filled me with a sense of mystery and magical potential. Before the escalating agitation of the great birds forced my mother and me to beat a prudent retreat, that nest made a lifetime impression.

Years later I began a decade of studying terns just down the beach from that same colony, and I visited the gulls regularly. When I began research toward my Ph.D. in ecology, I ran my boat each morning past the same island the gulls nested on and the very shoreline my mother had led me along.

For more than twenty years, I lived only about four miles from that gull colony, and each morning when I walked the mile from my home to the bay I saw that gull island. No one else in my professional world stayed in a single place for so long. Everyone went from home to college to graduate school to post docs to jobs.

I did most of these things, but just by chance, I never moved very far. During this lifetime in one place, I noticed changes in abundance of fish and other creatures. The fish I hunted for food and fun—striped bass, flounders, sea bass, sharks, marlin, tunas, plus sea turtles—all seemed in a continuous ebb tide of excessive catch and population decline. Fishermen I knew were grumbling, but virtually no one in the scientific community and not a single environmental group was talking about changes in fish populations.

Learned, sophisticated people, it seemed, just didn't stay in one place long enough to see changes over time. Funding agencies wanted results, not pointless, repetitive long-term monitoring studies. Other ecologists were obsessed with "hypothesis testing"—preferring to guess rather than patiently observe—a quicker route to "getting papers" and getting promotions.

But for the simple reason that I stayed put long enough to gain a place-based personal history, I witnessed the diminishment of my natural world. First it saddened me, then angered me, then outraged me to action. My approach to fishing changed and my career as scientist took a different direction. I wanted to tell everyone how drastic these changes had been. Personally witnessing history made me appreciate time's great orienting power. Time constantly transforms space. Like tide, it waits for no one.

Why the Past Is Important

Everything is on the way to becoming different, but in nature conservation, the past is the only rational guide to a better future. This is not true in medicine or electrical engineering or communications, where the past offers little insight on future developments. But we have diminished every realm of nature—forests, fishes, corals, climate—so thoroughly that almost no controls are left for comparison. The past must often become the control site.

Control sites are important. In tropical and subtropical seas, the U.S.-owned uninhabited Northwestern Hawaiian Islands and Palmyra Atoll are among few control sites left. Recent studies compared relative weight or biomass of big versus small fishes between the main Hawaiian Islands, the remote Northwest Hawaiian Islands, and Palmyra. The results: the weight of big, carnivorous fishes was only 3 percent of the entire fish community around the main Hawaiian Islands, but was 54 percent in the remote Northwest Hawaiian Islands, and even more around Palmyra Atoll. Even greater differences have been found when scientists surveyed the extremely remote Kingman atoll in the Line Islands—here, top predators comprised 85 percent of reef fish biomass. I've been to many of these places and the difference is profoundly striking—and a bit scary because of the abundance of big sharks. My book *Eye of the Albatross* relates my impressions of the amazing numbers of tuna and sharks around Midway Atoll—closed to fishing for half a century.

On the basis of the main Hawaiian Islands alone, no living person could have described the changes and no hypothesis could have been tested. So what of the rest of the world? We have no untouched Newfoundland to compare with the one we've fished for centuries.

The past is our only marker, orienting us in a trackless sea to the receding coast of our origins. Nature has no hope in the absence of history. But wringing information out of the past is problematic because scientists generally weren't around to document what was happening. Yet I want to ask whether ecologists overestimate this difficulty, insisting on standards of proof higher than necessary to get at the truth.

In other fields people seem to have less trouble accepting historical writings and authoritative anecdote. No one seems skeptical about what Europeans wore in the fifteenth century, or what their farm animals were like, or how Christopher Columbus's ships were built, though that information didn't come from scientists' clipboards.

So why does it seem unsatisfactory and unconvincing when we read Ferdinand Columbus's description that "in those twenty leagues, the sea was thick with turtles so numerous it seemed the ships would run aground on them and were as if bathing in them." Bathing in turtles? Surely, that can't be accurate!

We accept as credible Francisco Pizzaro's description of contact with the Incas but view as untrustworthy or even dismissible the notion of Caribbean turtles so locally dense during the 1600s that one Edward Long wrote, "It is affirmed that vessels which have lost their latitude in hazy weather have steered entirely by the noise which these creatures make in swimming." Both are equally anecdotal, yet even I will admit more skepticism about nonscientists' natural-history observation. I wonder why this is, and whether there is really proper justification for it.

If we are going to dismiss the writing of eyewitnesses, we should have better reason than the unconscious assumption that the world started on our first day of graduate school.

Many ecologists observe that the great drawback of historical information is that it was not scientifically, systematically collected. This is generally true, but not absolutely true. The scientific method is the most powerful toolkit, but it is not the only systematic way of acquiring information. For example, astronomy and the study of plate tectonics lack the formal scientific method of hypothesis, experimentation, and control groups. Yet these are sciences because science is the systematic pursuit of true facts that characterize existence.

Those explorers, traders, ship captains, and merchants who wrote the observations that historians now study were not scientists. Yet they were systematically pursuing something. They were motivated to keep track, to record observations. In an era before widespread professional science, their pursuit of new knowledge was often protoscientific. We can easily see a systematic approach in the record keeping of ship captains like Captain Robert Fitzroy of the *Beagle* (who kept better records of the origin of finch specimens than his young naturalist Charles Darwin) and Captain Charles Scammon, who meticulously documented the gray whales he nearly exterminated.

Science studies the messy, noisy, nonquantified world around us, selects things to focus on, and polishes its observations until the truth begins to shine through. Scientific observation, thought, discourse, and analysis can also be applied to material gathered and written by nonscientists. Historians are trained to think this way, yet ecologists seem *unduly* suspicious.

Yes, we *should* be suspicious of all sources of information. The world abounds in Trojan horses. Whaling and fishing captains often falsified their logbooks (especially as restrictions grew) and this has come to light spectacularly in several cases. Explorers had motive to hyperbolize their discoveries so their next expeditions could be funded. (Even scientists, despite greater efforts toward objectivity, can also be unconsciously affected by the funding imperative.) But not all historical information suggests great abundance. Some records speak of scarcity, and this further suggests reliability.

So scientists ought to at least be curious and not dismissive about the writings of seafarers who were dependable enough for investors of the day, competent enough to find their way to the edge of the earth and back, and thorough enough to systematically exterminate much of what they found. They got the first glimpses of a world that was round and whole. They were professionals on the leading edge of their time, and not all of them were lying or exaggerating. Their information was reliable enough to open fishing and whaling and sealing industries and trade routes. Their businesses required accurate accounting of the animals they killed. They had their stubborn superstitions, but we have our scientific orthodoxies. If we could meet today, we might understand each other all too well.

New attempts to look at the history of whale, turtle, and fish populations have ignited fierce criticism. The new analyses—each using very different techniques and timescales—have their strengths and weaknesses. Experience suggests that some are likely wrong, others likely right.

Let's make sure we debate the flaws and merits, not the notions. Let's make sure we are at least open to well-supported alternate ideas. Some people seem threatened by the very possibility that explorers were credible eyewitnesses or that removing a few million whales, a few million turtles, and billions of long-lived fishes might have changed the ocean.

Historical analyses in ecology are so unusual they have all the earmarks of big new ideas—shocking, heretical, minority-held. Let's bear in mind, as we shred new news with our learned criticisms, that many breakthrough ideas in science are at first attacked. Remember that geologists spent half the last century resisting the idea that the surface of Earth floats on moving plates. I myself independently discovered in third grade that South America looks like it once fit up against Africa—only to be assured that I was wrong. Had I had this insight fifty years earlier, resisted my teachers, and withstood a withering career of derisive criticism, I might have contributed something significant to geology.

We need many more controversial retrospective analyses along all available lines of investigation—historical, genetic, ecological—and they need to be as good as they can be, improved by competitive debate and by contributions of differing schools of thought. There is excellent information that is still low fruit, waiting to be analyzed. In 2003 Ram Myers and Boris Worm showed this in their analyses of existing, accessible data sets that had previously been ignored. Japan's longline data is almost certainly the longest-running ocean monitoring program in the world, but was overlooked until their study even by people who knew of its existence. Yet only by knowing, for instance, that 90 percent of the big fish are gone can we voice an informed recovery goal or an argument over what is sustainable. Be aware that the forces of evil are working the other side of the street. Tuna industry consultants and lobbyists removed the first half of the data set from Atlantic bluefin assessments because it showed a clear relationship between abundance of spawning fish and spawning success.

The past is more important because we are losing any sense of it. Our society shows profound disdain for history. Many cultures venerate elders and classic stories, but we are obsessed with modernism, unconscious of our roots. Animals, plants, habitats, and human cultures vanish. Even the memories of them are disappearing.

Why? To a significant extent our culture obliterates history and dismisses elders because no one owns history, and elders aren't lucrative corporate customers. Merchandising and fashion, thus advertising and the vehicles of advertising—magazines, TV, movies, the Internet—virtually ignore mature people, who often have sense enough not to buy what they don't need.

To younger generations who don't see the scam, perspective, wisdom, and history—even recent popular history—are practically invisible. A friend of mine who teaches maritime studies tells me that neither of his two teenage children has heard of Jacques Cousteau. Without the past we are disinherited, flying blind, naïve and vulnerable.

The past is important because the tragedy of the commons exists not just in place but also in time. Understanding what options and natural wealth have been robbed from us must motivate us to work against the robber barons now desecrating the great commons of time, sucking up the future, and stealing our descendants' natural endowment and compromising their well-being. Coral reefs, ice ecosystems, climate stability—great bites of nature have been degraded at the expense of future potential. Replacement cost will be measured in generations of lives over decades and centuries.

Society needs to get oriented. The past is important because the more we learn of it the better we might safeguard future options. A profoundly ahistorical society lives only in the present in the most dangerous way possible: consuming natural capital, while freezing and reversing progress on human rights and dignity. Those of us with a sense of history comprise a minority. These are very serious problems. To me the only correct response is moral outrage.

Good News

Now, the good news: the shifting baselines problem is not one of absence of data as much as absence of analysis and lack of communication. And both problems are improving. No students were surveying Native Americans' cod catches four thousand years ago, yet we have very informative reconstructions from Indian middens. No British or American scientists counted turtles in the Cayman Islands in the 1700s, yet we have very detailed population range estimates based on trade records and nesting beach observations. No Canadian observers with clipboards and hard hats were aboard fishing boats in 1900, yet we have a very informative historical analysis. Happily, this list is growing.

We should fill in the blanks with a sense of mission. Nothing is more precious than the rich, biologically diverse world that is our rightful heritage. That great tree of life contained all our future options. We have been handed a tree of life pruned and truncated. We must not hand the next generation a bush. To prevent further destruction, we need both science and the orienting power of history.

The world is in an anti-scientific mood. On Darwin's birthday in 2009 only 39 percent of Americans believed that evolution occurs, and the world is being set aflame by religious, political, and corporate ideologies. Yet, science is the anti-ideology: the search for truth. Just as martyrs died for religion and liberty, they also died for imagining new scientific truths. It is foe to cynicism. It is a threat to demagoguery, ignorance, political corruption, and corporate recklessness. Anyone fortunate enough to have received scientific training inherits a great moral imperative to spread the power of knowledge. This imperative should propel every one of us.

The past asks our trust. In exchange for that trust, it will help us focus our field of targets; it will inspire us toward healing and sober us with the parameters of plausibility. And even if the data are spotty and not quantified and we can learn only the broad generalizations and merely half of what we burn to know, that's still reason enough to seek, to dig, and to get to work.

Chapter 2

The "March of Folly" in Global Fisheries

U. Rashid Sumaila and Daniel Pauly

It has now become apparent to most interested parties that marine fisheries throughout the world are in serious trouble. Notably, the large, long-lived, and usually pricey fishes on top of marine food webs have declined by at least one order of magnitude relative to the period immediately following World War II, and global landings peaked in the late 1980s. In many coastal communities such as Newfoundland, fish stocks, which for centuries had supported vibrant fisheries, have collapsed with enormous economic and social costs. Similarly, in Ghana, West Africa, many locals no longer fish because their resource base has been depleted, largely by foreign industrial fisheries. This, in turn, has threatened the food security of local coastal communities.

Self-defeating practices are of course nothing new. In her famous book *The March of Folly*, historian Barbara Tuchman describes many instances of governments around the world pursuing policies contrary to their own self-interest and that of their people. The well-documented destruction of successive fisheries, and their supporting biodiversity and ecosystems, is a prime example of this folly. Current fisheries management policies led to the "march"—and endanger global food security. Yet there are measures that, if implemented, would overcome this pathology.

The Folly of Fisheries

Though fisheries scientists tend to be jaded about it, the public at large does not seem to realize the enormous impact fisheries have on marine ecosystems. Particularly misunderstood is the decline of fish biomass that occurred in the last decades, intensifying earlier trends initiated hundreds to thousands of years ago. This biomass decline, which affects predominantly the large, slow-growing, and expensive predatory fish on top of marine food webs, has consequences for marine ecosystems, as well as species facing extinction. Low biomasses also have dire economic consequences, including reduced incomes for the industry. The risk of recruitment failure (and subsequent quota reductions) increases because low biomasses tend to fluctuate more widely than high biomasses, which are usually composed of numerous and overlapping year classes.

Why is it so difficult to turn fisheries around? There are many explanations, some very technical, some more obvious. Here, Barbara Tuchman's "March of Folly" is a metaphor for the key features of the global crisis of fisheries. According to Tuchman, to qualify as "folly," a policy that is manifestly contrary to the interests of those pursuing it must meet three criteria. First, it must have been perceived as counterproductive when it was originally proposed, and not merely appear foolish by hindsight. Second, an alternative course of action must have been feasible; there is no point blaming anyone for pursuing a certain policy if there was no choice. And finally, the policy in question must be that of a group or institution, not an individual ruler or policymaker, and should persist beyond the (political) lifetime of individual members of that group. Tuchman went on to document the occurrence of folly throughout history, using a number of historical vignettes. Here, we will limit ourselves to documenting folly in fisheries, and then we will present potential remedies.

Regarding the first criterion, the definition of biological and economic overfishing that emerged in the mid-1950s was all that was needed to identify policies that would have conserved fisheries resources. The fisheries, however, continued to follow trajectories established much earlier: overexploitation was followed by the collapse of one species after the other and the subsequent expansion of the fishing fleets into more distant or deeper fishing grounds. All of this was subsidized with public funds, ultimately leading to a peak in global catches in the late 1980s with subsequent, ongoing declines (figure 2.1). Moreover, explicit warnings of the inevitable outcome of overfishing were consistently ignored. In the case of northern cod, coastal fishers in the Canadian Maritime Provinces, whose ancestors had

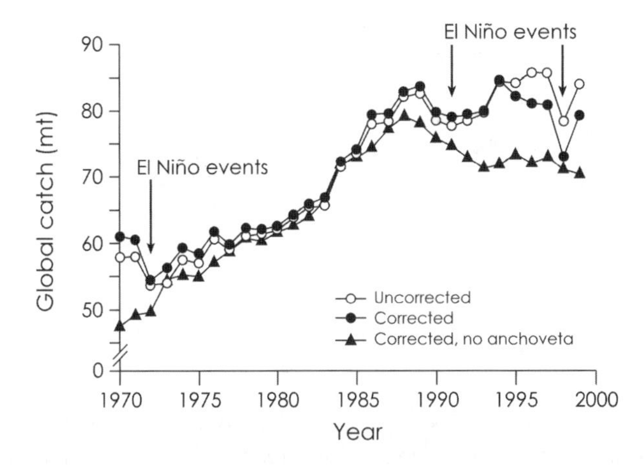

FIGURE 2.1. Global marine fisheries trends. Overall landings (*open circles*) are from the U.N. Food and Agricultural Organization (FAO) and suggest an increasing trend through the 1990s. Adjustment for overreporting by China, as proposed by Watson and Pauly in 2001, generates a decreasing trend (*solid circles*) though the 1990s. This decline, which continued in the 2000s, is even more apparent (*triangle, thick black line*) when the catch of the strongly fluctuating Peruvian anchoveta is omitted.

for centuries exploited cod in shallow waters using traps, argued against subsidizing the creation of a national trawler fleet after Canada declared an Exclusive Economic Zone (EEZ) and sent European deep-sea trawlers back home, but to no avail (figure 2.2).

The second criterion, the availability of an alternative course of action, was also met. In the 1950s and 1960s, fisheries biologists and economists advanced approaches to deal with the overcapitalization and overexploitation of fishery resources. They may not have been as sophisticated as some of the approaches currently proposed (and still not accepted). However, these measures would certainly have been more effective than the laissez-faire policy that was implemented and that led to a succession of stock collapses, culminating in declining global catches (figure 2.1). To return to the example of northern cod, the reasonable course of action would have been to let the cod stock, which European trawlers had decimated under open access, rebuild under the EEZ and to give the coastal fishers exclusive access to it (figure 2.2).

The third criterion—that the policy must be that of a group and be pursued for a long time—also clearly applies to fisheries. The overcapitalization and overexploitation of fisheries have been going on for a long time, well

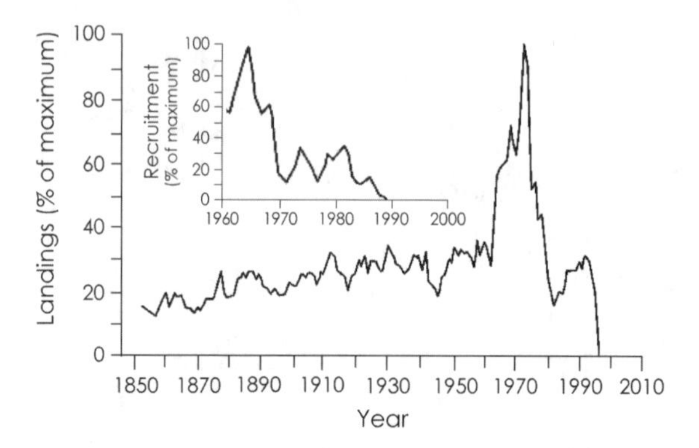

FIGURE 2.2. Time series of landings of northern cod off Newfoundland and Labrador, Canada. The slowly increasing and apparently sustainable catch from 1850 to the 1950s represents primarily the small-scale, inshore fishery that caught 150,000–250,000 tonnes per year over centuries (only the last century is shown here). The dramatic increase in the 1960s to the mid-1970s is due to foreign deep-sea trawlers that were succeeded by local Canadian trawlers in the 1980s. The trawl fleets drove catches up and the recruitment of young fish down (see time series in insert), resulting in the total collapse of the fishery.

beyond the political life of single governments all over the world. Indeed, fishing a resource down, then moving on to the next seems to have been the way fisheries have been run since time immemorial, certainly in the last decades. Moreover, we even forget about the previous abundance of resources we overexploit, and often end up claiming they never existed. This criterion undoubtedly applies to northern cod. Just consider the policies of successive Canadian governments, with the folly of excessive quota continuing even after a moratorium was declared in 1992.

What Foolish Policies Do

Foolish fishing policies have repeatedly resulted in damage to both ecological and human systems, affecting present generations and those to come.

Negative Ecological Effects

The folly of fisheries has resulted in declines in the biomass of most of the world's large predatory fishes. Myers and Worm demonstrated this for both bottom fish stocks and the large pelagic fishes of oceanic waters. Notably, the biomass of tuna and billfishes in various parts of the world ocean, as suggested by the catch per effort of the Japanese longline fishery, declined by a factor of 10 within ten to twenty years of being accessed by that fishery. Although this result was much disputed, at least for a few Pacific species, similar declines, using a very different, data- and computer-intensive approach, were shown for high-trophic-level neritic and oceanic fishes for the North Atlantic from 1900 to 2000, for Southeast Asia from 1960 to 2000, and for North West Africa from 1960 to 2000.

Fishing necessarily leads to declines in biomass, and some decline (to about half of unexploited biomass) is in fact necessary for a stock to generate a harvestable surplus yield. However, the declines mentioned above, those of northern cod (figure 2.2) and the multitude of stocks in the database assembled by the late Ram Myers, greatly exceed what was required to render the underlying stocks productive. Thus, as theory would have predicted, overall catches in the areas that suffered most from these declines in biomass began to stagnate, then to decline. For example, North Atlantic catches peaked in the mid-1970s and then went into continuous decline. This decline was masked by the geographic expansion of fisheries, that is, increasing landings of distant water fleets operating in places like North West Africa or the South West Atlantic, as well as by imports from even more distant areas such as the South Pacific.

However, this expansion proceeded at the rate of about 1 million square kilometers per year from 1950 to 1980, then accelerated three to four times in the 1980s. It has now reached its natural limits (figure 2.1). There was also a marked change in the composition of fisheries landings, which increasingly consist of smaller fishes from the lower part of the marine food web and reflect the scarcity of large, high-trophic-level fishes in the ecosystems (figure 2.3). This phenomenon, now widely known as "fishing down marine food webs" is far more pervasive than originally estimated (figure 2.4).

Negative Human Impacts

Historically, the answer to local overfishing has been to "move on," down the food web, toward deeper waters, and to other areas or regions of the world. Excess fishing effort by European vessels, for example, has traditionally been exported toward West Africa. Alder and Sumaila have demonstrated how this region of the world has attracted an increasing number of distant water fleets from western and eastern Europe, and from Asia

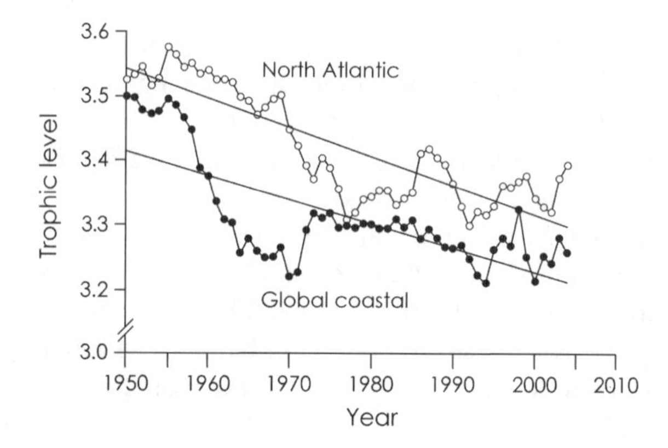

FIGURE 2.3. Decline of mean trophic level of fisheries landings reported by FAO. The data were mapped into 180,000 cells of 1/2 latitude/longitude degree according to the global information system (GIS) based procedure of Watson and colleagues in 2004. The two series shown here are for the North Atlantic (all waters north of 26° N) and for coastal waters globally, defined here as 200 kilometers out from coastlines.

FIGURE 2.4. What "fishing down marine food webs" means. Fisheries, after having removed the larger fishes at the top of marine food webs, must target fishes lower and lower down, including the prey and the juveniles of the larger fishes, which thus fail to recover. At the end, the fisheries target plankton species, including jellyfish. Bottom trawling compounds the problem by removing animals on the seafloor that offer shelter and food to the juveniles of numerous commercial fish species, thus further hampering their recovery.

between 1960 and 2000. The result has been steep declines in biomass in the waters off West Africa. West African countries received no real economic or social benefits for their depleted marine resources and suffered instead a serious decline in food security, as has been shown for Ghana's coastal fishing communities. Similarly, the collapse of cod stocks in Newfoundland in the early 1990s illustrates how an entire province in a developed country can suffer significant economic, social, and cultural losses because of foolish policies (figure 2.2). In addition to the huge economic and social costs imposed on the current generation, the destruction of marine resources will ultimately hamper the ability of future generations to meet their needs from marine ecosystems.

Some Reasons for Foolish Fisheries Policies

Foolish fisheries have their roots in fundamental failures in economics and governance, lack of well-defined catch rights, pernicious subsidies, technological progress, removal of market barriers associated with economic globalization, and shortsightedness in economic valuation.

Open Access and Common Property

The folly in fisheries is at least partially due to the open access and common property nature of fisheries. Fishing enterprises often operate without controls because catch rights—whether individual, communal, or state—are poorly defined, absent altogether, or not enforced. Open access or common property fisheries repeatedly lead to overcapitalization and overexploitation. This argument has led to various measures, especially at the international level, aimed at turning fisheries away from open access. A case in point is the 1982 U.N. Convention on the Law of the Sea (UNCLOS), which formally established the right of coastal nations to exploit, and the duty to protect, the marine resources within their 200-mile Exclusive Economic Zones. UNCLOS thus turned what used to be a global common into the property of coastal nations. However, UNCLOS did not solve the problem of "domestic open access," the problem of open access on the high seas, or the problems due to the transboundary or "shared" nature of some fishery resources. Thus, in many cases fisheries are still effectively open access, and hence the continuation of the "March of Folly."

Subsidies

Governments around the world continue to provide massive subsidies to the fishing sector, which intensify overcapitalization and overfishing. The worldwide amount of subsidies to the fishing sector is nearly $30 billion per year, much higher than the previous estimate used by the World Bank. This, when fisheries economists have shown again and again that subsidies enable otherwise bankrupt fishing fleets to maintain their pressure even on collapsed stocks.

Technological Progress and the Removal of Natural and Market Barriers to Overfishing

Over the last decades, new developments in fishing gear and sophisticated onboard electronics for fish finding, geopositioning, and the like have multiplied the fishing power of individual vessels, turning even small units into efficient fish-killing machines. Thus, fishing fleets have become powerful enough to overexploit essentially all stocks in the world, anywhere, any time of the year, thereby removing the last available "natural protection" afforded by depth or distance from shore. The loss of this natural protection, moreover, is not compensated for by the establishment of marine protected areas, which presently cover only about 0.7 percent of the area of the world's ocean. Technological improvements in preservation and transportation of fish products have also significantly increased the scope of international trade. This resulted in the removal of what may be described as market barriers to fishing: rather than being limited by their domestic market, the fisheries of various countries now have access to a global market with significantly higher demand for fish.

Shortsightedness in Economic Valuation

An important driver of the folly in fisheries is shortsightedness in valuation, stemming from the general human tendency to view what is closest to us as large and important, while discounting similar objects that are far away as small and less important. A similar tendency prevails with time, and it is operationalized by the economic concept of discounting. That is, values to be received in the future are reduced to their present equivalent value using a discount rate. In comparing the present values of policy alternatives, it is standard to discount net benefits that will accrue in the future compared to net benefits that can be achieved today. Cost-benefit analysis discounts streams of net benefits from a given project or policy alternative into a single number termed *net present value*. Thus, assumptions about the discount rate used in these time stream comparisons can have a huge impact on the apparent best policy or project. In particular, high discount rates favor the short-term policies that have led to unsustainable use of natural resources and particularly to global overfishing, as illustrated here for the recent collapse of northern cod off Newfoundland (figure 2.5).

Clearly, this widespread shortsightedness needs to be overcome. As Fearnside pointed out, decisions about the relative weight of short- versus long-term effects (in other words, the interests of current versus future generations) are a matter of policy rather than a scientific question. In principle, most of the world's countries favor a long-term approach. For example, the Magnuson-Stevens Fisheries Conservation and Management Act specifically requires that the interests of future generations be taken into account in the management of U.S. fisheries.

In practice, these decisions depend on the discount rate chosen to evaluate particular projects. Tol pointed out that the choice of discount rate and discounting approach is both empirical and ethical. It is empirical because people do make trade-offs between the present and the future in their daily

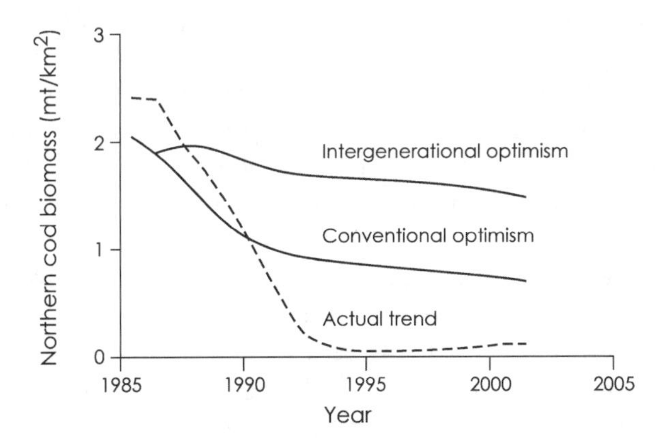

FIGURE 2.5. Comparison of cod biomass profiles under different management schemes and economic discounting. Note the large differences in the cod biomass resulting from the actual history of the fishery, characterized by utterly ineffective management, the biomass that would be left in the sea if conventional (and shortsighted) valuation and discounting had been used, and the large biomass that would be still be available if an intergenerational discounting approach had been applied.

decisions. It is at the same time ethical because the discount rate determines the balance of intergenerational allocation of goods and services. Tol suggested that neither the empirical nor the ethical should overrule the other. He also identified personal and social taste as factors in economic preferences. Since political choices reflect social taste, whereas personal taste reflects personal economic choices, people may prefer the use of lower discount rates to evaluate societal goals and objectives, even while using a higher rate for their personal decisions. After all, we know that at least some members of the current generation actually care about benefits to generations yet unborn. Of course, reversing the tendency to shortchange the future would stop the depletion of fishery resources and begin rebuilding depleted ecosystems for the benefit of future generations. Under these considerations, all generations should agree that overcoming shortsighted decision making would be the right thing to do.

Finding a Cure for Folly in Fisheries

There are four primary ways in which we can address and resolve the folly and collapse of fisheries.

Address Open Access, Common Property, and Subsidies.

To overcome the open access problem, we need more effective ownership structures at different levels, from the local to the national and beyond in the case of straddling stocks and high sea fisheries. The current literature on open access and common property tends to emphasize the privatization of fishing rights, although the revitalization of public ownership through lease or auction systems also appears promising and compatible with legal and cultural constraints in Western societies.

Additionally, we must forgo destructive subsidies. Subsidizing fisheries in developing countries is just as damaging as the grotesque subsidies in agriculture. Subsidization drives the folly of fisheries—without a stop the march goes on.

Reestablish Natural Protections by Creating Marine Protected Areas.

There is a clear need to reestablish the natural barriers to fishing that technological progress has removed by creating a global network of marine reserves. This would not only enable key stocks to recover, but also provide protection against assessment errors, acknowledged to be a common cause of collapses.

Engage in Aquaculture That Increases Rather Than Diminishes Food Protein Supplies.

Aquaculture has grown tremendously in the last decades, and is seen by many as the solution to food supply shortages. There are two main problems with this. First, the bulk of the reported increase in global aquaculture production consists of carp and closely related fish in China, and there are good reasons to assume that the overreporting besetting Chinese fisheries catch statistics also affects these aquaculture statistics. Second, in most other parts of the world, aquaculture growth is driven by increasing production of carnivorous fishes such as salmon, bass, and tuna, all of which consume more fish products than they themselves contribute. In terms of detrimental fisheries effects alone, this practice actually increases pressure on wild fish and removes cheap fish from developing country markets. Thus, aquaculture produces net fish for society mainly if it is limited to the farming of herbivorous fish such as tilapia and suspension-feeding invertebrates like mussels and clams.

Overcome the Tendency Toward Shortsightedness.

We absolutely need to deal with the still pervasive shortsightedness in valuation, as expressed by high discount rates. We must value benefits and adjust discounting practices in a manner that explicitly takes into account the interest of future generations as well as the present generation. This will only happen if natural resource scientists actively engage with and advise decision makers, making them and the public aware of the future consequences of considering only the economic, social, and political pressures of the here and now.

Acknowledgments

We thank Dr. Jeremy Jackson for his invitation to present our work at the KUU conference Marine Biodiversity: Using the Past to Inform the Future, held in San Diego, November 14–17, 2003, and for allowing us to summarize our presentations as a joint contribution. We are grateful to Dr. Reg Watson for extracting the data used for figure 2.3, and Ms. A. Atanacio for redrawing all our figures. Also, we acknowledge support from the Sea Around Us Project, initiated and funded by the Pew Charitable Trusts.

Chapter 3

If a Frond Falls in the Kelp Forest (does it make any sound?)

The Pew Oceans Report as a Case Study of Communicating Ocean Conservation

Randy Olson

Science is at the core of ocean conservation. Protecting nature and restoring it to a healthy state requires that we understand how nature works, and that means science. As a result, conservation is likely to be dominated by the scientific mindset, which is fine when it comes to the actual practice of science. But when that mindset begins spilling over into associated disciplines such as socioeconomics, policy, politics, and, most important, communication, the entire practice of conservation can become handicapped. That is what this chapter is about—understanding how "science think" can impede the effective communication of ocean conservation.

I'm going to use the first person in this chapter. I'm going to talk about communication, and if we know one thing about effective communicators, from Dostoyevsky to Mark Twain, it's that they generally speak in the first person. As in, "This is what happened to me." It's simply the most personal, and therefore the most powerful, voice.

Ideas vs. Events

In 2002 I was told that the Pew Oceans Commission Report was going to change America.

That's what I heard as Dr. Jeremy Jackson and I were beginning to assemble the basic ideas for our Shifting Baselines Ocean Media Project. I heard this from the communication directors of several major ocean conservation groups, and I heard it from Andy Goodman, communications consultant to the Environmental Defense Fund with whom I met for advice. He and others advised us to not bother exploring new ways to communicate ocean conservation because, as he put it, the Pew Report was poised to dominate the American media landscape. Any other communication effort would just end up being distracting "noise."

The Pew Oceans Report was the result of a three-year, $3 million study funded by the Pew Foundation to assess the condition of America's coastal oceans. It was billed as the most comprehensive assessment of U.S. oceans in thirty years—since the Stratton Commission Report of 1969. Other communications experts told me that the findings were so devastating that, once released, they would appear on the covers of *Time* and *Newsweek*, and would dominate the evening news for several days. From all directions I heard, "You better brace yourself, it's going to have a big impact."

There was some justification for these great expectations. After all, the 1969 Santa Barbara oil spill hit the mass media so hard it became a major catalyst in the birth of the modern American environmental movement. Furthermore, the decline of whales in the 1970s and syringes washing up on the beaches of Long Island in the 1980s put ocean issues on the covers of *Time* and *Newsweek*. So it's not like it hadn't happened before.

But the Pew Report was something different than a single, visually dramatic event. It was a thought rather than an act. Whether academics (and ocean conservationists) realize it or not, there *is* a difference between those two things. Communicating ideas is very different from communicating events.

A Coffee-less Press Conference

By the spring of 2003 we had raised some funds for Shifting Baselines and I had decided that what the world really needed was a television commercial that could call attention to ocean decline (a.k.a. "a public service announcement," though I hate to tarnish our work with a label that conjures up images of over-the-hill celebrities pouring their hearts out). My television director friend Jason Ensler and I had concocted a humorous sketch about the perils of lowered standards, and we thought it would be sufficiently entertaining to get plenty of free airtime. We were ready to film our

FIGURE 3.1. Maestro Jack Black speaks with CNN about his role in conducting the Bad Symphony for the Shifting Baselines television commercial.

commercial, and since the timing corresponded, we thought we might as well use the release of the Pew Oceans Report as part of our justification for filming.

I began contacting comic actors like Jack Black (figure 3.1) and, as part of my invitation to them, I said we were filming a television commercial to help call attention to the release of the Pew Oceans Report. To make sure this was okay, I contacted what remained of the production offices of the Pew Ocean Commission in Washington, D.C. I spoke with Justin Kenney, the fellow left in charge of staging a press conference to announce the release of the report. He was very warm and supportive of our efforts, welcoming any help to get publicity for the report.

I asked him about their strategy for communicating the findings. Would they be staging press junkets and assembling public events around the country? Holding town hall meetings with the commissioners to call attention to ocean decline? His reply: "Are you kidding? I barely have enough budget to afford coffee at the press conference."

A Textbook Example of "Science Think"

The plan was clear. The report was finished. And so was the budget for the Pew Ocean Commission. And this is what I refer to as "science think": It's the tendency to believe in the overwhelming power of information—to think that single ideas, data points, observations, factoids, or reports are what change the world.

The truth is, sometimes information alone is enough to change the world. And in a perfect world, it should be. But sadly, we don't live in a perfect world. We live in the United States.

So to make this long story very short, the release of the final report of the Pew Oceans Commission did not dominate the American media landscape. It was released on June 2, 2003. It did not make the cover of *Time* or *Newsweek*. It did not appear as the lead story on the evening news (though NBC did mention it a few days later). And it didn't even score the front page of the science-savvy *New York Times*, but instead ended up on page A-22 of that venerable publication.

Instead of going out with fanfare, the Pew Report landed in a pile of papers on anonymous desks.

The 9/11 Commission Report: The Exception That Proves the Rule

"It's just a piece of paper in the end." That's what the communications director of one major ocean conservation group warned me about the Pew Report. "The government puts out stacks of reports every day. It's hard for a report to get any notice in the media. Don't expect this one to do much." And clearly it didn't. But did that have to be the case? Not if you look at a similar report produced a year later.

The report was the final conclusions of the 9/11 Commission. On July 19, 2004, the *New York Times* reported on the unique approach the 9/11 Commission was taking to its task. Instead of falling victim to "science think" and just producing a report that would "speak for itself," the project was allocating equal resources to the "follow-through" stage. The Commission spent half its resources on making the report and the other half communicating what it had to say. As the *Times* reported:

> The lobbying effort would be a break with tradition, since blue-ribbon federal commissions often disband almost as soon as they have com-

pleted a final report, the members returning home from Washington and leaving the report to speak for itself.

You see that last sentence—that's exactly what happened with the Pew Oceans Commission. Only one of the commissioners, Roger Rufe, then Director of the Ocean Conservancy, went on the road to encourage others to put the report to work.

At the twenty-year anniversary of Surfrider Foundation celebration in October 2004, Warner Chabot of the Ocean Conservancy gave a talk about the Pew Final Report in which he said, "We've given you a brick, now go put it to use." Certainly a great line for inciting a revolution, but, alas, that was already more than a year after the release of the report and was simply a last-gasp plea to make up for the lack of brick throwing by ocean conservationists.

The Art of the Ad

What we're talking about here are two simple elements that can be broken down into the objective part of the process (the actual making of a product such as the Pew study) and the subjective part of the process (going out and communicating to other human beings about your product).

In the business world there is this thing called "advertising," which is basically that second part of the process. American businesspeople, more than anyone else in the world, know that it's not enough to just make a good product—you have to get out there and vigorously sell it.

Companies often see their sales increase significantly simply by changing their advertising strategy. As Ken Auletta detailed in an article in *The New Yorker* entitled "The New Pitch: Do Ads Still Work?" one guy at Aflac, saying the insurance company's name over and over again, realized it sounds like a duck. This advertising gimmick was enough to overhaul the insurer's entire image and double its sales in less than four years—without changing anything substantial about its product.

The need for the secondary, or follow-through, or just plain communication part of the process became broadly recognized in Hollywood with the 1977 release of *Star Wars*—the first blockbuster movie to tack on a gigantic marketing campaign. Industry skeptics back then ridiculed the waste of advertising funds at first, but as the box office erupted, thinking in Hollywood was changed forever.

Today it is simply accepted that studios will spend as much or more of

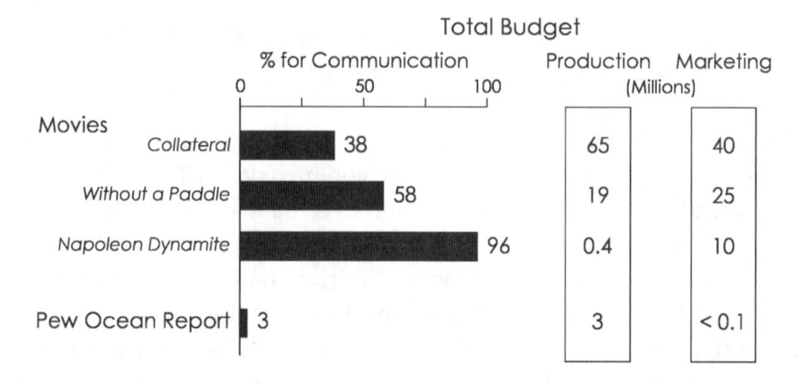

FIGURE 3.2. Financial allocations toward production of a product versus communicating to the public about the product (*Marketing*) for three Hollywood movies produced in the summer of 2003 (data obtained from *Box Office Figures*), and the Pew Ocean Report released in the same summer. Percentages represent how much of the budget went to communication. Any wonder why the Pew Ocean Report had so little impact on the American public?

their budget on advertising a movie as they do on actually making it. Figure 3.2 illustrates this point, showing the production versus marketing budgets for three representative movies in the same summer that the Pew Oceans Report was released. The most extreme example is *Napoleon Dynamite* ("Gosh, why'd ya have to pick that movie?"), which cost a mere $400,000 to make. A group of scientists would have acquired the movie, decided it was so brilliant that the quality would speak for itself, and saved $10 million by forgoing any ads and just putting it in theaters to catch fire on its own.

For some reason the studio execs didn't opt for that strategy. Instead they gambled $10 million on marketing (a scientist would say, "They could have made twenty-five more equivalent movies for that much money!"), and guess what came back to them at the box office? More than $50 million. So the makers of *Napoleon Dynamite*—a quirky outlier—ended up spending 96 percent of their budget to tell the public about their amazing product. Now let's take a look at what the Pew Ocean Commission did with the release of their final report. It spent roughly 3 percent of its budget on letting the public know what it had.

And there you have it. "Science think" at its finest: the belief that advertising/marketing/communication is a big waste of money. And the net result was a potentially important study had minimal impact because no one ever heard about it.

An alternative strategy would have been to spend only $1.5 million on a little shorter and simpler study that would have produced the same conclusions—that our oceans are in bad shape—and proposed what to do then, spend the other $1.5 million on a mass communication campaign to let people know this simple fact. So why didn't something like that happen? Let me use the first person again to give my perspective on it.

How About a Super Bowl Commercial

I gave a talk to a meeting of ocean scientists and conservationists in the fall of 2002. The meeting was dominated by an air of distress at how serious the problems in the world's oceans were becoming and the large gap that existed between this important information and the public's awareness of it. The organizers of the meeting had asked me to address the second problem and offer up suggestions on how to bridge the gap.

In my talk I threw caution to the wind by telling the group, "If you really think there is a major biodiversity crisis looming in the oceans that the public needs to know about, then I propose a simple solution to get the word out. I'll make a thirty-second television commercial for $50,000, then we'll spent the rest of the $2 million on buying a thirty-second time slot during the Super Bowl. Even if the commercial is terrible, every spot at the Super Bowl is so heavily scrutinized, the worst that will happen is the entire American public will scratch their heads and say things must be pretty bad in the oceans for someone to pay so much for that lousy commercial."

Was this a bad crazy idea? Definitely not! In February 2008 *ABC News* ran a segment posing the question of whether the presidential candidates should gamble $2 million on a Super Bowl commercial. The piece quoted the C.E.O. of GoDaddy.com, a Web hosting company. A few years earlier, he had taken an enormous gamble by spending half his company's assets on a single Super Bowl commercial: it showed a well-endowed woman testifying to Congress as her blouse falls partially off. The C.E.O. says, "Our decision was right as rain," as he recounts the gigantic surge in sales it brought them.

But my idea for a Super Bowl commercial prompted an instant argument among the audience members, with about two-thirds enthusiastically supporting me and a third saying I was a fool. The dispute climaxed with a Scandinavian fisheries biologist saying, "This is the *lowest* I have ever heard of science education stooping to."

And that pretty much sums up "science think." It's the tendency to believe that the silliness and irrationality of mass communication is a big waste of time and money. Except to brand their research, science organizations and foundations are rarely willing to support communications projects—much less *innovative* communications projects. It's a constant and ongoing problem.

Kindred Spirits: The *Turning the Tides* Report

I dug deeper into the world of ocean conservation and accumulated a longer list of frustrations with the lack of support for communication. But finally I was guided to a study that warmed my heart. I felt I was listening to the voice of kindred spirits. The study was titled *Turning the Tide: Charting a Course to Improve the Effectiveness of Public Advocacy for the Oceans*. It was funded by the Packard Foundation and it bravely hit the nail on the head about why ocean conservation is so ineffective.

Why was the study so powerful? It began by asking a very simple question: *Why is ocean conservation so ineffective compared to the major issue-oriented campaigns such as the anti-smoking lobby or the N.R.A.?* It found that each successful campaign was characterized by five key traits and that the ocean conservation movement lacked one of them: a realistic political strategy using lobbyists and real-world politics in Washington, D.C.

With a little further examination, the study's authors concluded that ocean conservationists are much more comfortable with and better at policy than politics. This matched everything I have experienced in working with the ocean conservation movement for nearly a decade. Ocean conservationists, scientists, and policy analysts know how to do research and make policy recommendations. But when it comes time to talk to the public to actually make things happen . . . well, the lack of success of the Pew Report says it all. It charted a course for sea change, but the nation didn't get the memo. Ocean conservationists are comfortable with showing up at the meeting, turning their backs to the public, mumbling to themselves, leaving a copy of their report on the table for people to look at if they want, then slipping out the door. Obviously this needs to change.

Throwaway Thinking in Our Throwaway Society

And now, in honor of Jennifer Jacquet and her "Guilty Planet" blog, it's time to add a little guilt to my message. Eco-conscious academics love to criticize careless consumption. But this "throwaway" mentality can infect our green friends, too.

We saw this in 2003 when talking with the communications directors of the major environmental groups. They held as sacred the basic belief that this year's message had better be different from last year's so you don't come off as stale. Instead of sticking to a single, consistent, well-crafted message, they were caught up in a never-ending cycle of creating new slogans, logos, and brochures. Is that really much different than single-use disposable plastics?

This same mentality was evident in the way the environmental community treated Al Gore's movie *An Inconvenient Truth*, released in 2006. By 2010 there was a widespread feeling that the movie was "old news," and it simply needed to be "let go," rather than "rehashed." But there's a difference between rehashing and follow-through. Just as the Pew Oceans Report need not be swept under the rug because it didn't have the desired impact, Gore's movie also needs a thorough postmortem. This is a critical aspect of follow-through—to assess what did and did not work about the project. That's how we get smarter over time. It's basically George Santayana's famous quote, "Those who cannot remember the past are condemned to repeat it." We can modify that a bit to say those who are unwilling to examine the past are doomed to keep releasing the same report year after year.

And So, Sadly, History Repeats Itself in 2009

In 2009 the Obama administration released what it called a "landmark" report on global climate change. The summary report of the United States Global Change Program was touted as "the most comprehensive and thorough report of its kind." Similar to the Pew Report, it was launched with a single press conference hosted by the President's Science Advisor, John Holdren—a video of which is on the U.S. Global Change Research Program website. But by the spring of 2010 I found myself at a talk at M.I.T. in which the speaker complained about how incredibly thorough the report was, and yet . . . nobody ever heard about it.

A Perfect World with Perfect Oceans

In 2003 Daniel Pauly, the individual most responsible for coining the term *shifting baselines*, published a book titled *In a Perfect Ocean*. For a perfect ocean to exist there would have to be a perfect conservation movement to maintain and defend it. And part of that perfect conservation movement would involve perfect mass communication between the people in charge of protecting the oceans and the people who want to use them. So what would such perfect ocean conservation communication be like?

The perfect movement would make "arouse, fulfill, and follow through" its universal slogan. All three parts of the process are essential. If the public isn't interested, everything downstream is of little value. Involvement has to begin with motivation, and most people are motivated not by facts and figures, but through human elements like emotion, passion, and humor. Once the mass audience is motivated to play an active role, feels ownership of this resource, wants to defend it, and reaches out for information, then its needs can be met through effective messaging. It's as simple as "arouse and fulfill." Once the audience's demands are fulfilled, the movement must follow through by evaluating the impact of its messaging, and reinventing and repackaging when necessary. Madonna has known this for three decades.

At its core, the perfect conservation movement succeeds by creating a voice that people want to hear. People respond positively to human voices. Even the most overused celebrity spokesperson is still better than a cold, lifeless statement from an organization. But reaching the mass audience in today's saturated media markets requires a new element—diversity. There was a time when a single spokesperson like Jacques Cousteau could reach the hearts and minds of the entire planet. Those days vanished in the 1980s with the information explosion. Today, individual demographics tend to "narrowcast" and listen only for voices with which they identify—voices similar to their own. A perfect conservation movement would appreciate the need for a diversity of voices and styles from all ethnicities and income levels.

And lastly, the perfect conservation movement would know that, above all else, communication is about people, which means that to effectively communicate any issue related to science or nature, in the end, you must find its "human face." It is always there, no matter how obscure the subject. And despite all the noise of today's information society, it is still the most powerful means of communicating with other human beings.

PART II

Anchovies and Sardines

> In the morning when the sardine fleet has made a catch, the purse-seiners waddle heavily into the bay blowing their whistles . . . the cannery whistles scream and all over the town men and women scramble into their clothes and come running down to the Row to go to work. . . . The whole street rumbles and groans and screams and rattles while the silver rivers of fish pour in out of the boats and the boats rise higher and higher in the water until they are empty . . . until the last fish is cleaned and cut and cooked and canned.
>
> —*John Steinbeck, Cannery Row*

Those were the good old days on John Steinbeck's Cannery Row when the California sardine fishery was booming in a confluence of industrial and natural abundance that seemed inexhaustible. Then the oceans changed. The fish and the stink were gone. Biological oceanographers call what happened "a regime shift"—a fundamental change in oceanographic conditions that was bad for the fish, bad for the processing plants, and bad for the people living on the margins. Now, the canneries have become high-priced commercial real estate for shops catering to affluent tourists, and the seedy and romantic characters of Steinbeck's world live on only in his books.

In his historical review fisheries scientist Alec MacCall examines the sardine-anchovy puzzle. The collapse of the California sardine fishery shortly after World War II generated the California Cooperative Oceanic Fisheries Investigations (CalCOFI) program, an ongoing, collaborative effort to develop a truly systematic fisheries science as a basis for sound management. A major participant in this work for more than thirty years, MacCall recounts its high ideals, conflicting agendas, dead ends, and lasting achievements, and reminds us of its pervasive influence in the marine sciences. Principal concepts and analytical tools, originally developed to investigate anchovy and sardine populations, have since become fundamental to fisheries management, marine biology, ocean science, and ecology in general. Discovering the link between the fluctuating abundance of these small pelagic fishes and decadal hemispheric climate cycles remains a scientific triumph with increasing significance in the era of global warming.

Using the Peruvian anchoveta fishery as their focal point, David Field and colleagues explore questions of uncertainly. What is known, unknown but knowable, and simply unknowable? In particular, they stress how uncertainty—including uncertainty caused by differences in interpretation—promotes mismanagement of fisheries and eventual human misery, such as occurred in factory towns along the Peruvian coast in the 1970s. Some of this uncertainty lies in the biological variability within the life cycles and migratory ranges of small pelagic fishes, still poorly known. More lies in environmental variability over differing spatial and temporal scales. Finally, Field and colleagues remind us that ecosystem damage from overfishing worldwide coincided with increased burning of fossil fuels and that the effects are very likely synergistic. Global warming has greatly increased uncertainty about the future of these fisheries.

Both chapters illustrate how scientific uncertainty, erroneously contrasted with certainty of opinion, has been used as a whipping post to prevent the implementation of rational fisheries regulation. While the negative statistical correlation between sardine and anchovy abundance may be weak, the human effects of the rise and collapse of the sardine fishery in California and the subsequent rise and collapse of the anchovy fishery in Peru are by now thoroughly convincing. Fundamental to this breakthrough was the pioneering application of paleoceanographic time-series data to expand the baseline for the fisheries beyond what was possible with conventional biological and climate measurements alone. Millennial time series, assembled from disparate sources and carefully calibrated, yielded solutions to previously intractable biological and ecological problems because they expanded conventional ecological frames of reference. This achievement heralded a slow but steady paradigm shift toward greater interdisciplinary collaboration. Despite remaining uncertainties, there is still room to improve management based upon our new understanding of the interactions between oceanographic forcing and fishing.

Chapter 4

The Sardine-Anchovy Puzzle

ALEC D. MACCALL

Sardine (*Sardinops* spp.) and anchovy (*Engraulis* spp.) populations around the world have exhibited extreme fluctuations, often varying a thousand-fold in abundance from one decade to the next, accompanied by economic boom-and-bust cycles that have become legendary. In nearly every case, fortunes are made during times of abundance, not only by the fishing and processing industries, but also by secondary industries such as poultry ranching and fish rearing—industries made possible by convenient large quantities of inexpensive, high-protein animal food. Yet the prosperity typically lasts for little more than a decade, and suddenly the fish stocks mysteriously disappear. In some cases, alternative fisheries are eventually developed. However, in nearly every instance of stock collapse, the social and economic damage is severe. Once prosperous fishing communities become ghost towns: processing plants are boarded up, equipment is sold, and large fleets of fishing vessels slowly rust away.

In response to this puzzle, biologists and oceanographers have conducted major research programs off California, Peru and Chile, Japan, and South Africa, but the answer has been remarkably elusive. Much of the work has been done in the California Current, the site of perhaps the largest, and certainly the longest, fishery-oceanographic research program ever undertaken.

Historical Review

During the 1930s and 1940s, the Pacific sardine (*Sardinops sagax caerulea*) supported one of the largest fisheries in the world, with annual catches exceeding 600,000 metric tons (mt) and fishing fleets active from Mexico to Canada. The collapse of this fishery in the late 1940s and early 1950s was a landmark event in fishery science and biological oceanography. Although a few other major world fisheries had disappeared previously, loss of the sardine fishery was one of the first to be viewed as a subject for large-scale scientific investigation, in this case centering on the debate as to whether the decline of the resource was due to overfishing or to natural causes, and consequently, what if anything could be done to rebuild the fishery. At the end of the 1940s, the resource was clearly in decline, and the industry was threatened by restrictive fishing regulations being proposed by California's Department of Fish and Game (CDFG). In response to this threat, the fishing industry underwrote the creation of an ambitious multiagency scientific program, the California Cooperative Oceanic Fisheries Investigations (CalCOFI), a program that continues to the present day. Its members included Scripps Institution of Oceanography (SIO), the federal government's Fish and Wildlife Service (FWS), and the CDFG, among others. The program was originally funded by the fishing industry through a self-imposed tax on fish landings and was overseen by the industry-controlled Marine Research Committee (MRC).

CalCOFI embodied the modern concept of science-based fishery management that was emerging rapidly in North America and Europe. However, the science itself was not yet firmly established, especially with regard to the basis of sustainable harvests. There was a bitter disagreement between CDFG scientists, who contended that overfishing was the cause of the decline, and the more lettered SIO and FWS scientists, who contended that a temporary period of adverse environment was at fault, implying that fishing pressure need not be reduced. Thus, the sardine debate mirrored the contemporary ecological debate regarding density-dependent or density-independent control of animal populations. The dark side of the CalCOFI debate was that the SIO scientists, who were the major recipients of the new funding, knew that the environmentally based CalCOFI program could not be justified if the problem was found to be overfishing. As McEvoy described it in 1986, "Through the first decade of its existence, then, the MRC-CalCOFI project perpetuated a finely tuned stalemate between government agencies competing for funds and influence, while the industry that oversaw it squeezed out what life remained in the sardine fishery."

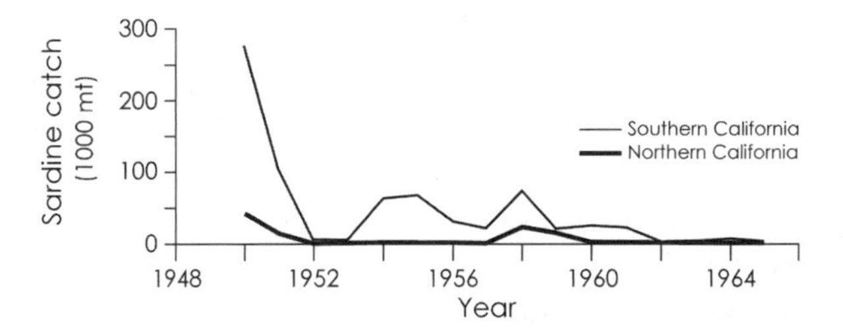

FIGURE 4.1. California sardine landings during the early years of California Cooperative Oceanic Fisheries Investigations (CalCOFI).

By 1950 and the beginning of CalCOFI, sardine catches had already declined to half the 600,000 mt level typical of ten years earlier, and in the 1952 and 1953 seasons no sardines could be found in California (figure 4.1). Although sardines reappeared in 1954, by the mid-1950s it was apparent that the resource was not returning to its former distribution and productivity. However, in 1957 through 1959, California experienced a strong El Niño that was accompanied by a resurgence of sardines, including their reappearance in central California. This provided convincing proof that environmental conditions have a major influence on the fishery and revitalized the CalCOFI program. It also brought about the realization that El Niño is associated with the physics and biology of the entire Western Hemisphere, including the California Current, and not just the west coast of South America. Thus, global climatology entered the mix of disciplines relevant to CalCOFI and the sardine problem.

In the early 1960s, Garth Murphy conducted a comprehensive study of sardine demography, during which he invented the now-standard fishery stock assessment tool of Virtual Population Analysis. Murphy concluded that overfishing was the primary cause of the decline in the resource—"Fishing rates applied to the population lowered reproduction to an extent that decline was inevitable."—and that reproductive failure in 1949 and 1950 precipitated the collapse of the stock. Murphy was not able to identify any specific environmental influence associated with the pattern of reproductive successes and failures.

Murphy also concluded that during the 1950s the northern anchovy (*Engraulis mordax*) stock had grown to a magnitude similar to the original sardine biomass. He argued that anchovies had ecologically replaced the sardines and were competitively preventing a sardine recovery. On that premise, Murphy, together with John Isaacs and others, promoted development of an experimental anchovy fishery designed to reduce competition with sardines. Although a small anchovy fishery did finally emerge in the 1970s, it was not economically viable. The proposed fishery also was politically unpopular: The recreational fishing sector was militantly opposed to a large anchovy fishery, fearing a reprise of the sardine collapse due to lack of management control and loss of a critical link in the food chain needed to support an abundance of sport fish.

Both the recreational fishing sector's concern with food chain relationships and Murphy's proposed experiment, based on classical ecological concepts of interspecific competition and species replacement, marked the entry of multispecies ecological theory into CalCOFI and the sardine debate, presaging modern calls for "ecosystem management." Murphy's experiment was never conducted, but it is worth imagining a scenario in which deliberate anchovy overfishing was enthusiastically attempted beginning in the late 1960s: Sardine abundance would have started increasing in the late 1970s (because we know it did so anyway), giving the impression that the experiment had worked! In hindsight, we also now know that Murphy's anchovy biomass estimate of 5,000,000 mt was severely in error. Using a much better technology to calibrate larval abundances, Lo and Methot later estimated that anchovy spawning biomass in the late 1960s had actually been less than 500,000 mt—a stock size that should be compared with proposed annual anchovy harvests of 200,000 to 1,000,000 mt. Thus, the experiment could indeed have depleted the entire anchovy biomass in two or three years—more quickly than the triennial CalCOFI larval surveys would have been able to detect, realizing the worst fears of the recreational fishing sector. Both sides of the argument would have been able to claim victory in the debate, and yet fishery managers would have been none the wiser for it.

A primitive form of ecosystem management was also being considered in the Peruvian system, where Schaefer treated the guano birds (mostly cormorants, *Phalacrocorax bougainvillii*) as a competing source of mortality in the Peruvian anchoveta (*Engraulis ringens*) fishery. Combined abundance of the three seabird species had declined from 28 million individuals in 1955 to 4 million in the late 1960s. Schaefer calculated that loss of the birds had increased sustainable yield of anchoveta by roughly 2,000,000 mt. He concluded that any recovery of the bird population would require a reduction in allowable harvest by the fishery and that at the current guano bird abundance, "the annual anchoveta catch can be maintained indefinitely at 9.3 million metric tons." His only statement in favor of the birds was that the birds should not be eliminated entirely.

Within three years of Schaefer's analysis, the Peruvian anchoveta fishery had collapsed, and the seabird population had also declined to fewer than one million individuals. The immediate cause of the declines was the strong El Niño of 1972, which caused reproductive failure of both fish and birds and also caused unusually high vulnerability to an already intense fishery. However, the anchoveta's lack of response to subsequent reductions in fishing effort was puzzling. Meanwhile, sardine (*Sardinops sagax sagax*) abundance was increasing rapidly in Peru and Chile. On the other side of the Pacific Ocean, *Sardinops melanostictus* was increasing even faster in Japan.

In the early 1970s, Soutar and Isaacs developed a remarkable time series of prehistoric sardine and anchovy abundances based on fish scales preserved in southern California laminated anaerobic sediments. The 2,000-year paleosedimentary record, since refined by Baumgartner and colleagues (figure 4.2), indicated that unfished sardine abundances were highly variable off California, with occasional disappearances even in the absence of fishing. Once again, the fishery was absolved of responsibility for the disappearance of the resource: "Nor can the virtual absence of the sardine from the waters off Alta California be considered an unnatural circumstance." Lasker and MacCall later pointed out that Soutar and Isaacs' conclusion was not valid because the paleosedimentary data were incapable of resolving sardine abundances below about 700,000 mt, at which level zero scale counts become frequent. A biomass of 700,000 mt would still be considered relatively healthy, whereas recent sardine abundances were estimated to be below 10,000 mt. Another of Soutar and Isaacs' surprising findings was that there was no indication of anchovy-sardine alternations of

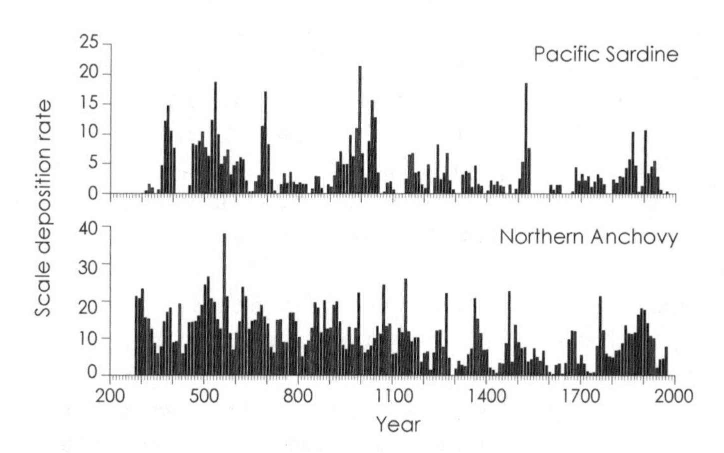

FIGURE 4.2. History of sardine and anchovy scale deposition rates (no./1,000 cm²/yr) in the Santa Barbara Basin, southern California. (Redrawn from Baumgartner et al. [1992].)

abundance in the paleosedimentary time series, despite scientific consensus that the two species were competitors.

The pattern of sardine fluctuations implied by the paleosedimentary record could not be reconciled with the conventional fisheries view of an approximately constant "reference" state of the resource corresponding to an unfished condition (i.e., carrying capacity, in ecological terms). At the 1973 CalCOFI Symposium, Isaacs formalized this concern:

> [T]here are probably a great number of possible regimes and abrupt discontinuities connecting them, flip-flops from one regime to another. . . . Sardines, for example, are either here or not here. . . . There are internal, interactive episodes locked into persistence, and one is entirely fooled if one takes one of these short intervals of a decade or so and decides there is some sort of simple probability associated with it . . . organisms must respond to more than just fluctuations around some optimum condition. . . . Fluctuations of populations must be related to these very large alternations of conditions.

This was the origin of the terms *regime* and *regime shift*, which have recently become common keywords in fisheries and oceanographic publications.

Lasker and MacCall examined the widths of the scales from Soutar and Isaacs' study. Based on the relationship between scale width and fish size, Lasker and MacCall concluded that the average anchovy was 54 percent heavier during periods when sardine scale deposition was low. While this superficially seemed to be further evidence of sardine-anchovy competition, they concluded that the difference was a probably coincidental response of anchovy growth to the (unknown) environmental conditions that influenced sardine abundance. Specifically, Mais showed that in the late 1970s a sudden reduction in the average size of anchovies had occurred in southern California. Sardine abundance, while showing initial signs of increase, was still much too low to have influenced the anchovy's food supply.

In the early 1980s it was becoming apparent that sardine fluctuations occurred synchronously on a worldwide scale. Major sardine fisheries had developed almost simultaneously in Peru-Chile and in Japan, where the sardine population expanded to occupy the Kuroshio Current extension and a large portion of the northwest Pacific Ocean. A new fishery had developed in the Gulf of California, and sardine abundance was increasing in the California Current. In 1983, the Food and Agriculture Organization addressed this and related fishery issues by convening an Expert Consultation to Examine Changes in Abundance and Species Composition of Neritic Fish Resources. At this meeting, the worldwide synchrony of sardine fluctuations was captured vividly in a figure presented by Kawasaki, reproduced here in figure 4.3. Although the synchrony in Kawasaki's figure is exaggerated by combining the Gulf of California sardine fishery with the California Current fishery (they are separate stocks), the strength and contrast of the relationship suggested that the underlying mechanism should easily be discovered. Moreover, at the same meeting, Parrish and colleagues presented a comprehensive study of the fishery-oceanographic mechanisms governing fish recruitment in eastern boundary currents worldwide, where most of the sardine stocks occur. It appeared that all of the pieces of the puzzle were now in hand. A breakthrough seemed to be imminent, generating major symposia in Capetown, South Africa, and Vigo, Spain, in 1986 and Sendai, Japan, in 1989. Yet the answer proved elusive.

In the 1980s, CalCOFI oceanographers and biologists were beginning to notice that the California Current ecosystem was behaving differently than it had during the early years of the program. Sea surface temperatures were consistently warmer than those seen in the 1950s and 1960s, and sardine abundance was clearly increasing, among a wide variety of other physical and biological phenomena. Venrick and colleagues were the first to fully identify the scale of the change as a regime shift: "We postulate that these environmental fluctuations have resulted in significant long-term changes in the carrying capacity of the [Central North Pacific] epipelagic system. . . . We need to re-evaluate both the assumption of steady state . . .

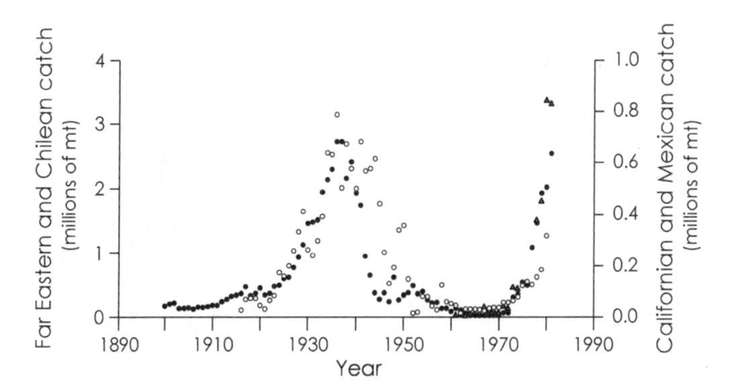

FIGURE 4.3. Kawasaki's demonstration of synchronous worldwide fluctuations in sardine populations. Solid circles are Far Eastern sardine, open circles are Californian and Mexican sardine, and triangles are Chilean sardine. (Redrawn from Kawasaki [1983].)

and our studies of community structure and dynamics." The existence of a major shift in northeastern Pacific climate ca. 1976 achieved popular recognition with publication of a multivariate study in the meteorological literature by Trenberth in 1990. The 1976 climate shift provided the first directly observed phenomenon that could account for the kinds of fluctuations seen in the paleosedimentary record. It also helped recast the sardine puzzle from the conventional fisheries "recruitment problem" of understanding year-to-year fluctuations to a new "regime problem" that was concerned with coherent worldwide decadal scale variability of anchovies and sardines. The fishery and oceanographic research community was slow to embrace the regime idea, and it was about ten years later that the concept of "regimes" became widely accepted (figure 4.4).

The California sardine fishery had been closed by a legislated moratorium in 1974, with the provision that a fishery could be resumed if sardine abundance recovered to at least 20,000 short tons (18,144 mt). By the mid-1980s signs of increase were unmistakable. In 1985, estimated abundance had reached this level and a small fishery was allowed. Sardine abundance was closely monitored, emphasizing an ichthyoplankton-based spawning area survey in the southern California Bight.

Changes occurred in several worldwide sardine-anchovy fisheries following 1988. Recruitment to the Japanese sardine fishery declined suddenly, and under intense fishing pressure, the resource biomass declined by 95 percent between 1988 and 1992. Although the turning point in Peru

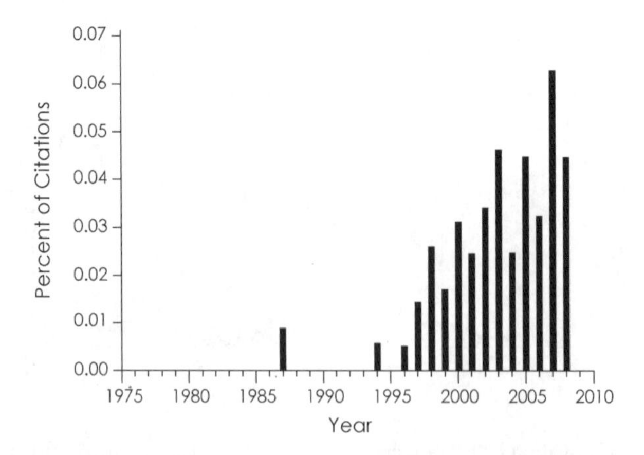

FIGURE 4.4. Fraction of peer-reviewed publications containing the keywords *regime* together with *climate* and *fish* or *fisheries* in the *Aquatic Sciences and Fisheries Abstracts* database.

was not as clear as that in Japan, Peruvian sardine fishery catches declined by 99 percent during the 1990s, while anchoveta catches were returning to pre-1972 levels. South Africa experienced a post-1988 warming of the Benguela Current and sudden growth in its sardine population. Although a 1988 regime shift clearly occurred in those regions, no corresponding shift was apparent in the California Current where the existing warm conditions intensified.

Sardines continued to increase in abundance off California, and the southern California spawning area increased progressively from 1985 to 1991. The spawning area surveys were based on the conventional view of sardines as being a relatively nearshore species. Indeed, the historical fishery had been conducted in nearshore waters, and the early CalCOFI ichthyoplankton surveys also suggested a coastal affinity. This nearshore view was shattered in 1991 due to the chance discovery of large concentrations of Pacific sardines far offshore. While conducting exploratory trawling for jack mackerel (*Trachurus symmetricus*) in international waters more than 200 miles off the California coast, the Russian survey vessel *Novodrutsk* encountered surprising abundances of Pacific sardines as well as chub mackerel, a.k.a. Pacific mackerel (*Scomber japonicus*)—at the farthest edge of the range covered by standard CalCOFI surveys. Also, Pacific sardines reappeared in British Columbia in 1992 after a nearly forty-year absence. This was attributed to the increasing abundance of sardines off California and a northward shift in distribution due to the strong 1991–92 El Niño.

The 1990s saw major advances in understanding of interdecadal climate variability. Mantua and colleagues described the Pacific Decadal Oscillation (PDO), a pattern of low-frequency atmospheric variability in the north Pacific that is related to but is not identical with the previously known pattern of El Niño-Southern Oscillation (ENSO) that dominates low-frequency variability in the south and equatorial Pacific. It was apparent that fluctuations of sardine and anchovy stocks were related in some way to patterns in the ENSO and PDO, but the mechanism was unclear.

In the mid-1990s, environmentally explicit models of the stock and recruitment relationship were developed for both the Pacific sardine and the Japanese sardine. Previously, the effects of environmental conditions on reproductive rates could not be distinguished from the effects of parental abundance because of co-linearity in the data. However, post-1976 data from California and post-1988 data from Japan provided new information with contrasting environmental and abundance information that allowed analytical separation of those effects. Jacobson and MacCall found that for a given parental abundance, sardine recruits per spawner were about twice as high during favorable environmental conditions as they were during unfavorable conditions. Under favorable conditions, equilibrium yields could approach 1,000,000 mt, but during unfavorable conditions, there may be no sustainable yield to support any fishing whatsoever. Wada and Jacobson found that Japanese sardines achieved a remarkable twentyfold increase in recruitment during favorable environmental conditions, which also explains the rapid growth of the Japanese fishery after 1970. Also, the "switch" from a favorable to an unfavorable state is abrupt; the virtual cessation of sardine reproduction explains the rapid depletion of Japanese sardines by fishing after 1988.

There have been several recent attempts to synthesize the information on sardine and anchovy fluctuations. Schwartzlose and colleagues presented a comparative study of low-frequency variability in sardine and anchovy systems around the world. McFarlane and colleagues examined a variety of hypothesized physical and biological mechanisms, but concluded that "the underlying mechanisms . . . have yet to be identified." Chavez and colleagues reviewed the characteristics of regime shifts from an oceanographic perspective, but the puzzle was not solved: "It remains unclear why sardines increase off Japan when local waters cool and become more productive, whereas they increase off California and Peru when those regions warm and become less productive."

Overfishing or Environment?

From the inception of CalCOFI, every few years a scientific study should be published that indicated a strong environmental influence on California's sardine fluctuations. These were carefully written publications, and the authors tended to be cautious in their interpretations. However, in each case the popular press was eager to announce that this most recent study "finally" proves that natural events rather than overfishing were the cause of the historic collapse of California's sardine fishery. Apparently, scientific publications that confirm overfishing as the primary cause of the sardine collapse are of little public interest, but those that support a societal denial of responsibility are newsworthy. It is ironic that the same conflict of interest that Scripps Institution of Oceanography had in the 1950s (and which it outgrew) survives to the present day in the "newsworthiness" of sardine analyses reported by the popular press.

Of course, the question of whether the historical sardine collapse was due to overfishing or to adverse environmental conditions is not posed correctly. It fails to recognize that the collapse was due to both causes, working in concert. This relationship can be clarified by some simple calculations based on historical fishery catches and population estimates given by MacCall in 1979. The environment shifted to an unfavorable condition sometime during the 1940s. Between 1950 and 1965, the fishery landed a total catch of 928,000 mt. During that period, the population size declined by 777,000 mt, implying that net production of sardines during that period was 151,000 mt. Thus, during those sixteen years, average sustainable harvest was about 9,400 mt per year, in contrast to actual average harvests of 58,000 mt per year. Approximately 16 percent of the actual catch was sustainable, whereas the remaining 84 percent was "mined" permanently from the resource. The average fishing mortality rate of 36 percent per year was very close to the natural mortality rate, which has long been considered to be a safe rule of thumb for fishery management. However, in hindsight, if only 16 percent of that total harvest was sustainable, the true sustainable fishing rate must have been only about 6 percent per year under the prevailing unfavorable environmental conditions. In view of the past fifty years of worldwide "expert opinion" in fisheries, this would have been an inconceivably low harvest rate to propose for a sardine fishery under any conditions. Only in hindsight do we know that the sustainable harvest rate was only one-sixth of the "safe" rule of thumb described above. A collapse of the sardine industry could not have been avoided.

A fishery management plan for California's sardine fishery was adopted in 1998. The harvest specifications under this plan are extraordinary in the history of fishery management. First of all, the plan establishes a minimum sardine biomass reserve of 150,000 mt, below which no harvest is authorized. Based on the stock and recruitment relationships described by Jacobson and MacCall in 1995, the allowable harvest consists of a temperature-dependent fraction of the biomass in excess of 150,000 mt. That fraction ranges from a maximum of 15 percent under favorable warm conditions, down to a minimum of 5 percent under unfavorable cold conditions. Because of the reserve biomass, realized total harvest rates will be below 5 percent under cold conditions and declining abundance. It remains to be seen how this management policy performs in future decades, and it cannot be evaluated fully until the sardine stock passes through the next cold period and emerges into the following warm period. The hundredth anniversary of the founding of CalCOFI should be an appropriate occasion to evaluate its success. Meanwhile, the post-1985 fishery on the Pacific sardine is still healthy after twenty-five years. No other sardine fishery in the world has lasted more than twenty years.

Chapter 5

Variations in Fisheries and Complex Ocean Environments

David B. Field, Francisco Chavez,
Carina B. Lange, and Paul E. Smith

Today, Peru's anchoveta fleet and fishmeal factories remain idle for most of the year. Although fish are abundant, large numbers of vessels resulting from overinvestment mean that seasonal quotas can be reached shortly after the fishing season opens (figure 5.1). After a brief period of intensive work, hundreds of fishers, dockworkers, and factory employees must search for jobs in whatever other employment may be available the rest of the year. However, these hardships pale in comparison with the social and economic upheavals that followed the anchoveta population collapse in the early 1970s, when coastal villages became virtual ghost towns. A generation later, as catch rebounded to nearly 10 million metric tons per year, reinvestment in the fishery soared (figure 5.2).

Since the 1950s, decades of highly abundant Peruvian anchoveta have spurred enormous investment in fisheries and processing plants, while decades of low catches have brought about the loss of industrial capacity, boats, and local employment. Similar boom-and-bust cycles have plagued sardine and anchovy fisheries in the California Current, Kuroshio Current, Benguela Current, and many other regions. In general, strong El Niño events result in lower anchoveta catches, but landings rebound quickly and investment in factories and vessels remains high (figure 5.2). During the 1960s, this resulted in industrial overcapitalization. Within a

FIGURE 5.1. *Top left*, The anchoveta fishing fleet of the Port of Callao, Peru, remains idle during a seasonal closure. *Top right*, Surrounded by the fleet, a fishing vessel sets a purse seine amid a school of anchoveta. *Middle left*, Fishmeal processing plants in Chimbote, Peru, remain idle during a seasonal closure. *Middle right*, An anchoveta seiner raises a full net. *Bottom left*, Recently caught anchoveta. *Bottom right*, Filleting anchoveta is time consuming, but when fresh fish are available, the result is tasty. (Photos by Erich Rienecker and David Field.)

decade, however, the anchoveta population had collapsed and towns along the Peruvian coast were devastated. Landings declined sharply and remained persistently low, and the Peruvian fishmeal industry remained moribund for almost twenty years. In 1984, a yearlong moratorium on anchoveta fishing followed a strong depression in biomass after the 1982–83 El Niño (figure 5.2). Then in particular, factories and fishing vessels subsisted on the less abundant populations of sardines.

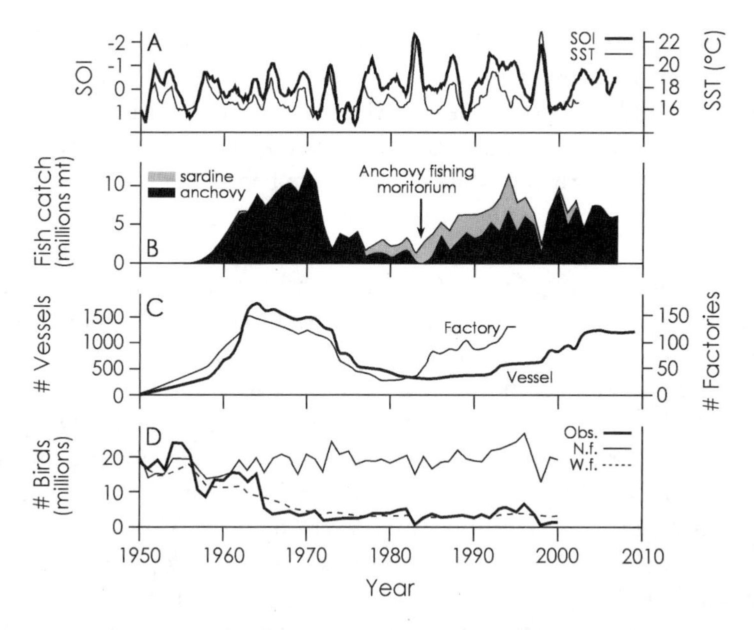

FIGURE 5.2. An example of the interaction between climate, fish catch, development of the fishing industry, and predators of small pelagic fishes. (A) Thirteen-month moving average of Southern Oscillation Index (SOI) and sea surface temperature (SST) at Puerto Chicama as indicators of El Niño events and ocean climate off Peru. (B) Peruvian catch records of anchoveta and sardine. (C) Number of anchoveta fishing vessels (industrial and artisanal) and fishmeal processing plants (Fréon et al. [2008]). (D) Ecosystem response of predators of small pelagic fish (Jahncke, Checkley Jr., and G.L. Hunt Jr. [2004]) indicated by actual estimates of guano birds from observations (*obs.*; *bold line*), a model prediction of guano bird abundance with no fishing included (*n.f.*; *solid line*), and a model prediction with fishing included (*w.f.*; *dotted line*).

Small pelagic fishes like anchovy and sardine have very high egg production rates and growth rates, and recruitment varies widely from success to failure. These factors make them susceptible to large fluctuations in abundance and exacerbate the boom-bust cycles of their fisheries. Their immense coastal populations depend on highly productive but complex and unstable boundary current ecosystems. The difficulty in determining how oceanic processes interact to affect recruitment, abundance, and distribution exemplifies the difficulty in distinguishing natural variability from overfishing, despite decades of intensive study, long-term records, and monitoring. Moreover, because small pelagics sustain higher trophic levels of fish, their dynamics are fundamental to entire coastal pelagic ecosystems.

Today, the Peruvian anchoveta fishery has rebounded once more, and once more the idle boats reflect overinvestment. The Institute of the Seas of Peru (IMARPE) closely monitors anchovy abundance and sets quotas that have helped maintain moderately high catch for years. Seasonal closures maintain a stock structure with sufficient numbers of large spawning fish, and this stabilizes the fishery. Smaller fish have much higher water content that decreases the quality and quantity of fishmeal. When catch is dominated by small fish, changing fishing locations or halting activity altogether for a few days or weeks could yield greater value per fish, but work habits are persistent and such restraint is rare. Unfortunately, business as usual precludes taking full advantage of the potential value of small pelagics as a resource.

Domestic costs due to ecosystem change have already been borne by the once powerful guano industry, which, in the 1950s, objected strenuously to the development of the anchoveta fishery because it threatened seabirds and guano production. This concern appears to have been justified by the precipitous decline in seabirds following the development of the anchoveta fishery in the 1960s and 1970s (figure 5.2D). Although guano mining is even less efficient than fishmeal production at converting protein-rich anchoveta into a valuable industrial product, using fishmeal to feed livestock and farmed fish such as salmon is enormously wasteful compared to direct human consumption. Unfortunately, markets for anchoveta as human food are very limited.

We now know that decadal changes in sardine and anchovy abundance may be related to large-scale changes in Pacific climate that persist for decades at a time. The development, collapse, and subsequent rebuilding of the Peruvian anchoveta fisheries in the 1990s graphically illustrate not only the interaction between natural variability and fisheries exploitation, but also how this interaction affects ecosystems and economies (figure 5.2). In this chapter we review aspects of the known, unknown, and unknowable regarding these highly productive, variable, and relatively well studied fisheries. One major emerging uncertainty is how climate change may drive ecosystems in unprecedented directions. We conclude with a discussion of some of the most important implications for future fisheries management.

The Known

The Efficacy of Fishing Restrictions

The magnitude and duration of a fisheries collapse, and its economic consequences, increase dramatically if restrictions are not imposed long before fishing becomes economically unviable. Unfortunately, an increase in catchability and price per fish is not uncommon in diminishing fisheries (e.g., figure 5.3A). In 1964, the California sardine population was approaching its lowest levels observed to date, when reports from swordfish spotter planes revealed the locations of major sardine populations offshore. Immediately they were targeted and heavily fished. Thus, driven by high demand and high price, catch per unit effort actually increased for a time. Three years later the sardine fishery collapsed and was shut down, just as the Peruvian anchoveta fishery was on the rise.

The same industrial participants who had presided over the collapse of the California sardine fishery shifted focus to the anchoveta fishery in Peru. Machinery and infrastructure went south. Fishery fleets and processing plants soon developed the capacity to extract several times the number of fish that existed in the wild (figure 5.2C). In 1972, evidence pointed to a

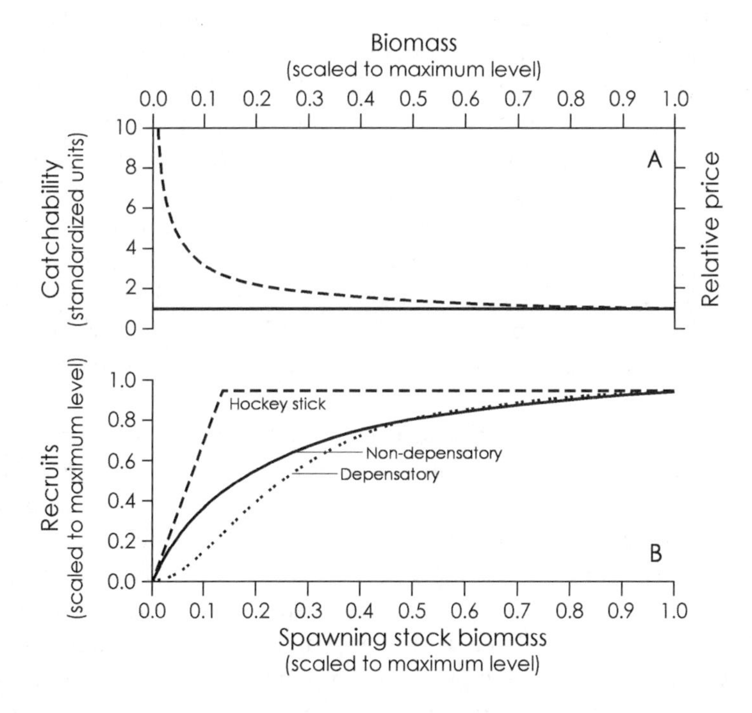

FIGURE 5.3. Models of response of fish populations to changes in adult biomass: (A) Changes in catchability and price per fish may remain constant (*solid line*) or increase at lower population sizes (*dashed line*). (B) Theoretical stock-recruit relationships showing different levels of density dependence. Spawning stock biomass and recruits are scales to their maximum levels, but these vary dramatically with environmental conditions. (Modified from Shertzer and Prager [2007]).

drop in recruitment the year before and declining biomass during the onset of the 1972–73 El Niño. Yet, scientific recommendations to reduce total catches were ignored. High catches justified the continued heavy fishing. Thus, the anchoveta population was soon decimated by a combination of overfishing and El Niño, which had concentrated the diminishing population near the coast (e.g., figure 5.4), giving the false impression of increased abundance. After the collapse, population growth and the abundance of anchovies failed to recover for decades (figure 5.2B).

Variations in Recruitment

How fish growth and recruitment respond to changing environmental conditions and preexisting abundance is critical to determining the effect of fishing on future population size, and current and future fish yield. The effect of fishing may be negligible if population growth shows negative density dependence (compensation) due to competition or cannibalism, or density independence because environmental factors dominate. Growth of juveniles may also be faster if adults are removed because competition for food is reduced. Thus, recruitment and growth may exceed the existing population size and support removal by fisheries. In the early years of the Peruvian anchoveta fishery, adults were removed at high rates under the premise that this encouraged growth of juveniles. However, poor recruitment in the early 1970s and concurrent removal of adult biomass resulted in a greatly diminished spawning biomass, which, along with less favorable oceanographic conditions, had an adverse effect on recruitment in future years.

It is widely believed that greater spawning biomass increases the probability of high recruitment when spawning biomass is low (figure 5.3B). Although actual recruitment data don't follow any model well, nearly all models assume that recruitment increases with greater population size and spawning (figure 5.3B), even more so when spawning aggregations are too small or habitat insufficient. The "hockey stick" model indicates that recruitment will vary positively with spawning biomass until a threshold is reached and the relationship breaks down (figure 5.3B). For example, sardine recruitment in the Kuroshio Current appears to be largely determined by environmental factors, and density dependence is important only at low population sizes. The population threshold may also vary with climatic conditions so that continued fishing below a critical population size may limit the potential future increase of the species, especially during

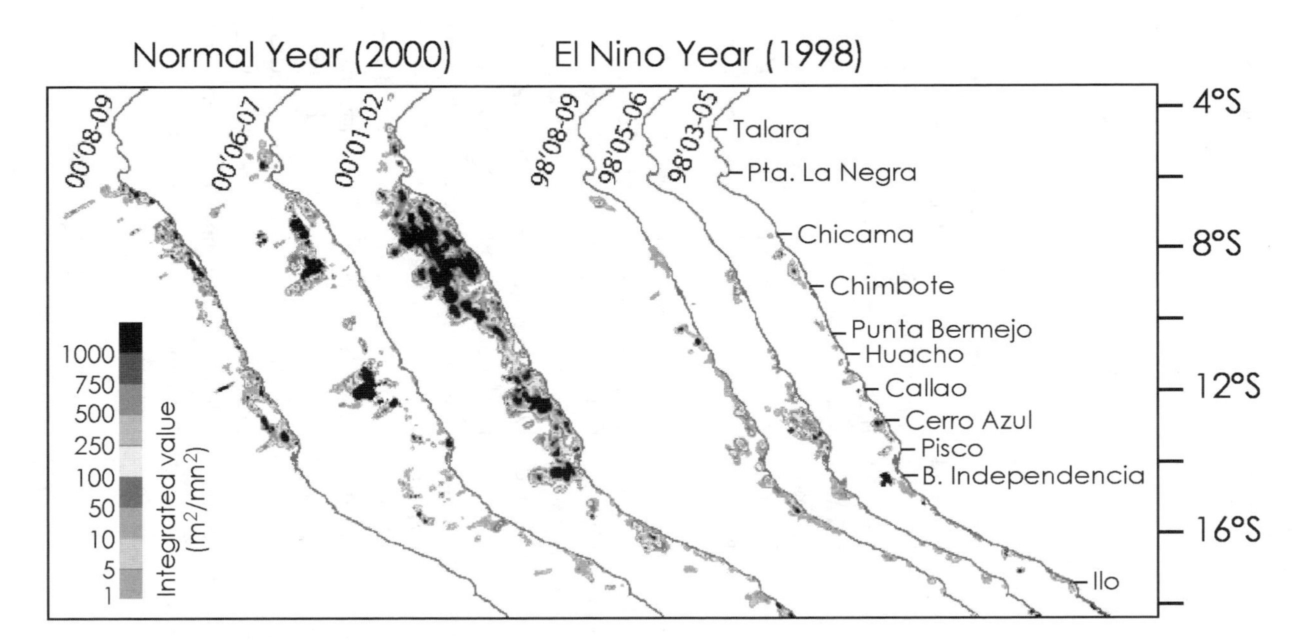

FIGURE 5.4. Spatial distribution of anchovy biomass along the coast of Peru as inferred from back-scattered acoustic energy for anchoveta spatial distribution surveys taken during different seasons (year and months) of a normal year (2000) and an El Niño year (1998). (Following Castillo et al. [2008] and Ñiquen and Bouchon [2004]).

unfavorable oceanographic conditions. Since variability in recruitment is always high and dependent upon environmental conditions, estimates of critical population size or maximum sustainable yield (MSY) are at best crude approximations.

The management plan for the California sardine is unusual in the incorporation of temperature variation as a factor in the regulation of the fishery. However, the relationship between temperature and recruitment is still not well understood. Chavez and colleagues have emphasized variations in productivity that are particularly evident between regions and with El Niño events, whereas Logerwell and Smith focused on mechanisms of larval recruitment related to retention and circulation determined by mesoscale eddies, and MacCall emphasized flow patterns on basin scales.

Very strong El Niño events off the Peruvian coast clearly reduce the abundance, growth, and productivity of the Peruvian anchoveta, but the cause of decadal-scale variations in abundance is obscure. The Southern Oscillation Index (SOI) was more strongly negative and waters warmer when sardines were more abundant than anchovies (figure 5.2), but anchovy recovery occurred during the early 1990s when the SOI had been negative for many consecutive years.

The California Cooperative Fisheries Investigations (CalCOFI) time series from the California Current exhibits only a weak relationship between zooplankton biomass and fish larvae from 1950 to 1970, suggesting that trophic links are not strong. Moreover, the abundance of sardines increased during the 1980s and 1990s when zooplankton decreased. However, the decline in zooplankton is largely attributed to gelatinous taxa, such as salps and doliolids, which are generally of little dietary importance to fish, whereas the most abundant species of copepods and euphausiids did not decrease. Sampling problems affect the determination of mechanisms that drive recruitment and growth because fish collections have primarily sampled fish eggs and larvae rather than juveniles, and many zooplankton taxa sampled are not preferred prey for sardines or anchovies. Nevertheless, the lack of clear trophic links on decadal timescales illustrates the uncertainty that remains regarding the environmental factors controlling population growth and recruitment, so management must proceed without a precise model of underlying mechanisms.

Variations in Fish Size, Type, and Distribution

The existence of different stocks or age structures of small pelagic fishes, distributed across large biogeographic regions, greatly complicates stock assessments and understanding of recruitment dynamics. There are well-documented changes in the distribution of small pelagics in the Kuroshio Current, California Current, and Peru-Chile Current. For example, as El Niño reduces the area of cool productive waters off the Peruvian coastline to narrow areas near the coast, anchoveta congregate in these coastal regions, move south toward Chilean waters (figure 5.4), and shift their vertical distribution deeper by tens of meters. The density and catchability of fish can increase near the coast during a strong El Niño, even while total abundance, recruitment, and growth are extremely reduced (figures 5.3 and 5.4). Large-scale spatial data on stocks are clearly essential if distorted estimates of population sizes are to be avoided and management becomes more successful.

Variability in the spatial distribution, age, size, and stock of a species sometimes results in local and regional differences in recruitment dynamics that greatly complicate relationships between future recruitment, stock size, and environmental variability. Different stocks may exist within a boundary current system, and migration can occur between regions and stocks. Anchoveta off Chile and southern Peru are generally considered to compose a different stock than that off central and northern Peru, but migrations occur between them. Older, larger sardines are found from northern California to British Columbia during population expansions and show a more wide-ranging swimming behavior. The presence of larger, older fish in the northern California Current may result in greater recruitment throughout the California Current since larger fish find food more easily, can produce more eggs, and spawn in a greater range of habitats. Sardine catches are more stable off southern and Baja California, which may be due to the existence of a separate stock. If considerable mixing occurs, fishing in one region may affect future catches in another region and they should be managed as a single stock. However, the mere existence of recognizably different stocks suggests that independent management strategies may be more appropriate for each stock.

Ecosystem Structure and Indicators

Management must also consider the importance of a target fish population to other components of the ecosystem. Small pelagics are important prey to seabirds and higher trophic level fishes including tunas and salmon, which are important to commercial and recreational fisheries. However, it is unclear how much higher trophic level fish depend on sardines and anchovies. If sardines or anchovies are unavailable, top predators may feed on other small pelagic species, undergo decreases in population growth, or change migrational patterns to feed in regions with greater availability of prey. Sardines and anchovies are traditionally considered such important baitfish for recreational fishers that the 1948 California Proposition Initiative 15 would have prohibited the use of purse seines for commercial fishing in some regions of the California Current in order to preserve sardines for the sports fishing and recreational industries. When anchovy were abundant but sardines absent in the 1960s and 1970s, recreational and commercial fishers of higher trophic level fish supported commercial catch limits of anchovy to maintain availability of small pelagics as food for these species. Apparently, competing interests divided the California fisheries along the trophic levels of their targeted species. The perception by fishers that fishing in the California Current is a zero-sum game has been with us for nearly a century.

Guano birds are historically the most conspicuous predators on Peruvian anchoveta. Populations of the guanay cormorant, Peruvian booby, and Peruvian pelican exceeded 20 million birds and consumed an estimated 1.3 to 2.1 million metric tons of anchoveta per year, prior to the inception of the fishery. Since anchovy live deeper and farther out of the range of diving birds when sea surface temperatures increase, the birds are strongly affected by prey reduction associated with El Niño events. However, the abundance of guano birds in Peru decreased dramatically with the development of the Peruvian anchoveta fishery and has not recovered along with the anchoveta (figures 5.2B, D). Populations of seabirds may provide the most accurate indicator of large-scale abundance of small pelagic fishes because they forage over much broader regions than the limited survey tracks of monitoring programs.

Historical Records

The large variations in small pelagic fishes on decadal timescales are a major problem for understanding patterns of variability because biological time series of abundance are too short. Sedimentary records of fish scale abundance and historical observations reveal several aspects of variability in population size prior to catch records and oceanographic surveys. The basic assumption for interpreting the sedimentary records is that shedding of fish scales and their consequential flux to the sediments accurately reflects the number of fish living in the water column, which seems to be well founded. While shifts in the spatial distribution of pelagic populations introduce uncertainties about the meaning of scale flux records from a localized core, much has been learned from sedimentary records.

The mere presence or absence of scales can be highly informative. The northern anchovy was not known to inhabit the Gulf of California when the species unexpectedly appeared in the sardine catch in the late 1980s. The initial interpretation was that fishing of sardines had created a new niche for anchovies that would otherwise be outcompeted. However, analysis of sedimentary records indicated that anchovy populations had persisted in the Gulf of California previously and that their recent occurrence was part of the natural variability of the ecosystem.

Time series in abundance of fish scales in sediments from the Santa Barbara Basin are well known for indicating great variability in abundance of sardines, anchovies, and hake for two thousand years before the advent of commercial fishing. Less emphasized is the fact that anchovy and sardine scales vary together at times, generally on centennial timescales, and vary out of phase at other timescales, generally on decadal timescales. Environmental variability affects each species differently and in ways that do not depend on the abundance of other species. Thus, abundance is not necessarily driven by competition or climatic conditions that always favor one species over another. However, Lasker and MacCall found that the size of anchovy scales is considerably smaller when sardine scales are abundant, suggesting that competition for resources may occur when both species are abundant, causing a reduction in average size and weight of anchovies. These observations caution against fishing one species to increase abundance of another.

Anthropogenic Activity versus Natural Variability

The recent downward trend in zooplankton displacement volume in the California Current was associated with a large warming trend; both of these trends reached their extreme levels during the 1997–98 El Niño. Cooler conditions in the ensuing La Niña event of 1999 and following years resulted in zooplankton biomass and temperatures nearer the long-term average although not back to levels observed in earlier decades. Such reversals in temporal trends epitomize the difficulty in distinguishing a long-term trend possibly due to global warming from natural variability in modern records, such as CalCOFI surveys that began in 1950.

However, analyses of planktonic foraminifera from a sediment core record from Santa Barbara Basin clearly demonstrate that, in about 1925, the abundance of tropical and subtropical species began to increase substantially above the variability typical in preceding centuries, whereas temperate to polar species showed no trend or decrease (figure 5.5). Principal component analysis based on the abundances of these species shows a strong association with sea surface temperature within the twentieth century and indicates a substantial warming trend beginning before most ocean and climate time series began. CalCOFI hydrographic surveys make up one of the longest records of ocean climate, but the entire time series lies entirely within a period of unusually high abundance for tropical and subtropical species. The 1999 La Niña event brought nearly record low temperatures in many instrumental records, but is only a short-lived minor variation within the longer-term warming trend observed in the core records (figure 5.5).

In addition to the twentieth-century warming trend, the assemblages of planktonic foraminifera species in the Santa Barbara Basin after the mid-

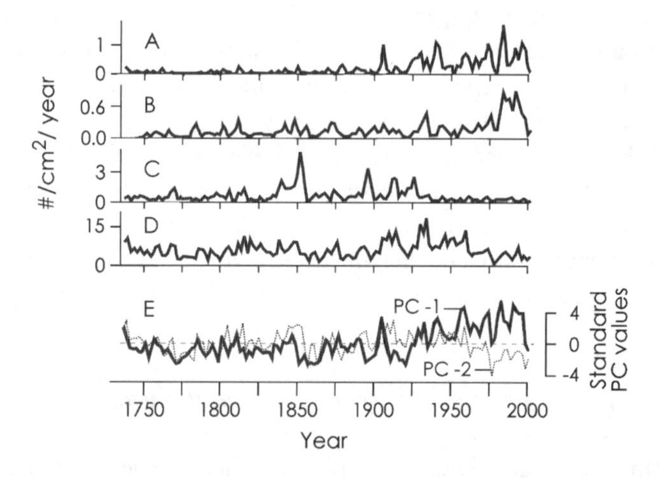

FIGURE 5.5. Temporal variations of several species of planktonic foraminifera from two-year sampling intervals of Santa Barbara Basin sediments illustrating how much of the twentieth century has an assemblage of species that is not typical of prior centuries. Species that are of tropical and subtropical origin and show trends of increasing abundances are (A) the combined abundances of *Globigerinella calida*, *Globoturborotalita rubescens*, *Globigerinita glutinata*, *Globigerinella siphonifera*, and *Globigerinella digitata*, and (B) *Orbulina universa*. Species that are of temperate and subpolar geographic affinities and show no trend are (C) *Neogloboquadrina pachyderma* (sin.), and (D) *Turborotalita quinqueloba*. (E) Temporal variations of a principal component (PC) analysis of the major species that show unusual separation of PC-1 from PC-2 during the late twentieth century. (Modified from Field et al. [2006]).

1970s are strikingly different from assemblages that existed during earlier decadal variations (figure 5.5). Coral records from the tropical Pacific also indicate that temperatures following the mid-1970s shift in the North Pacific exceeded prior variations throughout the last 150–1,000 years. It is well established that higher sea surface temperatures in the tropical Pacific drove an intensification of the Aleutian Low Pressure system, resulting in physical and biological shifts in the North Pacific in the mid-1970s. Taken together, these records indicate that the change of the mid-1970s in the North Pacific is distinct from most natural variations and is most likely explained by an influence of greenhouse gasses on ocean climate. The combination of models and instrumental data further indicates that an influence of greenhouse gasses on the ocean and atmosphere was noteworthy in the late twentieth century and will continue in the future.

The anthropogenic influence on ocean climate in the twentieth century coincides with anthropogenic removal of many mammals and fishes at higher trophic levels, which can have subsequent cascades to lower levels. Many populations of sea otters, sea lions, northern fur seals, and whales were severely depleted and elephant seals were commercially extinct by the late nineteenth or early twentieth century. The recovery of many of these populations of marine mammals in the twentieth century coincided with the intensification of many other commercial fisheries. Studies of the relationship between climate and marine populations in the North Pacific are derived from an atypical period of intense anthropogenic activity that has influenced many aspects of both ocean climate and marine populations. As the trend of climate change in the California Current is expected to continue, albeit with substantial variability superimposed, the changes in ecosystem structure are moving to even more unknown and perhaps unknowable states.

The Unknown and the Unknowable

We still have much to learn about the physical processes and trophic dynamics that affect growth and recruitment in small pelagic fish populations. Variations in the mortality rates of larvae and juveniles strongly affect total recruitment but have not been well sampled compared to eggs. Likewise, additional spatial and temporal sampling would help to determine the relative influence of variations in productivity, mesoscale eddies, basin-scale flow, and other processes that affect population recruitment and growth. Some of these measurements could be made from networks of buoys at a more limited cost, but the expense and effort would still be enormous. Time series of oceanographic observations from the California Current are among the longest available from the oceans but are still shorter than large decadal variations. Obviously, we need longer time series, but how long do they need to be and do we have the time to wait and see?

We need to improve our ability to predict future oceanographic conditions and their consequences for marine populations based on ocean-atmosphere-ecosystem models. Complementary to such predictive models are good reconstructions of the past. New records of fish scales, productivity, and climate are being developed from laminated sedimentary records from sites off Peru and Chile. The discovery of laminated sediments at Effingham Inlet has raised expectations for finding additional sites in the fjord regions of the northeast Pacific and southern Chile. Historical, archeological, and geochemical observations provide additional information on population distribution and abundance preceding commercial catch records.

Other high-resolution paleoclimate indicators provide additional information on climate variability affecting fish populations that may improve understanding of the specific links between marine populations and climatic changes. To do this, it is necessary to distinguish between synoptic patterns and time-averaged patterns of atmospheric and climate variability. Are marine populations controlled more by lower frequency, mean, or accumulated environmental conditions, or by the activity of higher amplitude, but perhaps infrequent, extreme synoptic events? For example, event-scale variability in upwelling or eddy formation may be more important for sardines and anchovies than the average wind strength or flow of a given year. Better time series of paleoclimate and fish scales could contribute greatly to this basic question.

But we need to ask how much does any of this new knowledge about climate variability and fish populations contribute to better management? Complete and definitive answers are probably unattainable. More important, in almost every case the collapse of a fishery was obvious beforehand from data on declining stocks. Time after time, resistance from fishers and policymakers, combined with questions and uncertainties about the data, prevented action from being taken. We will never have enough information to cast aside all the uncertainties or doubts about the relations of fish stocks to the condition of marine ecosystems and climate variability. The federal mandate to calculate and assign to different fisheries a Maximum Sustainable Yield is meaningless unless environmental variability is accounted for, but this is never fully possible. Moreover, changing climate and anthropogenic change will inevitably introduce new uncertainties. The sardine fishery in the California Current is among the best studied in the world, but we still lack a solid mechanistic understanding of the linkage of fish populations to their environment. Management decisions must be made without knowing all the answers, and to pretend otherwise is folly.

The ocean is an extremely complex environment, and fundamentally important climatic and ecosystem processes operate at different spatial and temporal scales within and among different oceanographic regimes. Climate change and the extirpation of species add whole new dimensions of uncertainty to the mix. High-latitude ecosystems such as the Bering Sea may be more impacted by climate change than the California Current due to greater magnitude of rising temperatures and loss of sea ice. In the California Current, species may simply shift their ranges northward, although new combinations of species may occur.

Extant species have experienced dramatic changes in climate associated with glacial-interglacial cycles and very rapid warming events, but future temperatures will likely exceed those of the past several million years. A warmer climate will be coupled with other anthropogenic activities and stresses. Reduction in the abundance of sardines or anchovies to the point of commercial extinction increases the risk of extinction of their predators such as larger fish or birds. Commercial extinction is unlikely to put small pelagic fishes at risk of biological extinction since numbers of individuals may still be in the millions and reproductive rates are enormous compared to marine mammals or sea turtles. Nevertheless, mitochondrial DNA of sardines and anchovies shows shallow haplotype diversity indicative of severe population bottlenecks or founder events in the recent past. The appearance and disappearance of northern anchovies in the Gulf of California support the genetic view that populations expand into new habitats, and stocks and genetic diversity can virtually disappear. With the combined pressures of fishing, ecosystem reorganization, and future climate changes there is considerable uncertainty in the future of the Anthropocene. However, we have learned some basic principles and lessons.

Lessons for the Future

Small pelagic fishes such as sardines and anchovies exhibit extreme variability in recruitment, biomass, and distribution patterns that help to sustain great fisheries but also make them vulnerable to overfishing. Despite great uncertainties in the details of population regulation, several lessons for the future are clear.

1. *Fishing a population in decline inhibits its ability to recover*. Fishing alters stock structure in ways that may reduce future recruitment, and density-dependent recruitment at low population sizes indicates that populations that are greatly reduced should not be further fished. Fishing populations to depletion, particularly during unfavorable climatic conditions, reduces their ability to recover during subsequent more favorable conditions.

2. *Diversity in fish populations is beneficial to the health of the stock*. Diversity in the age and size structure of a population and broad geographic distribution are beneficial to recruitment. Older and larger fish often have greater lipid reserves, span larger habitat ranges, and produce more eggs, which increase the probability of future recruitment.

3. *High fish densities do not imply healthy populations*. Extreme spatial variability in populations can substantially distort perceptions of abundance. Pockets of high densities of anchoveta appear nearshore off Peru during El Niño events when growth and reproduction are severely limited. These aggregations often indicate a contraction of a diminishing population to the nearshore, not a healthy population spread across a large spatial area.

4. *Highly dynamic ecosystems require precautionary management*. Sustainability is an obscure concept in dynamic systems like the California Current where the only certainty is change. Populations are always expanding or contracting in response to varying oceanic conditions on top of the effects of fishing, so adaptive management is essential on yearly or seasonal timescales. Adaptability is all the more important in the face of climate change.

5. *Manage fisheries with an ecosystem view*. The abundance of sardines and anchovies is of vital importance to many species of fish, seabirds, and marine mammals as well as other industries. The extreme depression of seabirds in relation to overfishing of anchoveta is obvious (figure 5.2D), but effects on larger predatory fishes are not as well understood. Abundance of euphasiids, birds, and mammals, as well as fish oil content, may be useful indicators of the state of the ecosystem in the presence of fishing. However, interactions between species are complex, and fluctuations in

abundance of species dependent on the same resource are not strongly predictable. Thus, culling one species in order to increase the abundance of another, as proposed for seals versus cod and whales versus fish, will almost certainly have unexpected and unwanted consequences.

6. *Unchecked exploitation leads to grim social consequences*. Overcapitalization and the tragedy of the commons plague fisheries. Business practices favor current exploitation over future potential, as well as the current and future potential of other fisheries and other industries such as tourism. There is a clear need to limit current fishing capacity and address incentives that threaten the long-term interest of the fishery itself and the marine environment. However, fisheries are only easy to close when there are too few fish to sustain the fishers.

How Can We Use These Lessons?

The oceans are complex. Lessons learned from complex scientific inquiries may be simple, but their application to sustainable management is greatly complicated, not least by human nature and the particular legal, economic, and social values associated with certain species. Now seasonal closures protect Peruvian anchoveta, and the California sardine management plan permits flexible harvest guidelines by factoring in temperature and population to determine harvestable biomass. However, West Coast groundfish have been in sharp decline off California for years. Groundfish are subject to environmental variability monitored by CalCOFI, but they have different life-history characteristics that are not the focus of sampling. Groundfish live longer than sardines and exhibit different recruitment patterns and relationships with the environment. Tighter fishing restrictions on groundfish are now in place, but there is considerable outrage from fishers who complain that the science is uncertain—this in one of the best-studied ocean environments in the world! There will never be enough information to address every uncertainty.

The fisheries crisis demands that we not end with the common refrain that "more research is needed." We must apply learned lessons now, and design institutions and strategies that are flexible enough to incorporate new information emerging from ongoing research that includes not only the best science, but innovative social, legal, economic, and cultural approaches as well. Forty years ago, the biggest investment into fisheries was for vessels and the greatest exploration was in developing new fisheries. Today, we need to invest in and explore new ways to ensure the future sustainability of fisheries resources. One way would be to make the economic value of fish reflect its ecosystem value. Lower trophic level fish such as sardines and anchovies could be used directly for human consumption without the 90 percent loss in energy that occurs when they are fed to tuna and salmon in fish pens, or fed as fishmeal to chickens and hogs. Ironically, sardines were once considered a delicacy in California, and their value as a healthy food is high. Cultural values change through time, and it will take a positive shift in values to ensure that ocean resources and marine ecosystems are restored to health and managed sustainably.

Acknowledgments

We thank Renato Salvatteci, Miguel Ñiquen, and Jaime Jahncke for their help, guidance, discussions, and data on the Peruvian anchoveta fishery and ecosystem. Alec MacCall, John C. Field, and Pierre Fréon provided helpful comments and discussions on the manuscript.

PART III

Cod

The story of cod is dramatically different from that of sardines and anchovies because physical oceanographic signals are weak and overfishing apparent at every stage. This story is also related differently than others in this book. The following chapters on cod differ in length, in structure,.and in the way evidence is presented because the principle authors are historians, not scientists. They take great pains to explain historical epistemology, which they would not have done in a text written for colleagues in their own field. Often, the different disciplines ascribe similar concepts with radically different meanings. These chapters help to clarify the differences between historical and scientific perspectives, in hope of developing a common understanding.

Unlike the cumulative body of sardine-anchovy research, now half a century old, the work communicated by Jeff Bolster, Karen Alexander, and Bill Leavenworth is relatively new. The estimate of cod biomass on the Scotian Shelf in 1852 was first presented at the Scripps conference in 2003. Like the sardine-anchovy research, it involved intense collaboration, but investigators expanded the interdisciplinary paradigm well beyond science. The historians discovered, extracted, analyzed, assembled, and linked data sets in formats that marine scientists could then employ in population models. More than a recapitulation of previous work, this chapter provides the rationale for historical analysis. It reintegrates scientific results and historical analysis. Thus, the biomass estimate derives relevance and power from the testimony of long-dead fishermen.

The Scotian Shelf study refutes long-held myths about the limitless abundance of the oceans and the inability of traditional fishers to harm regional fish populations. It proves that reliable historical data sets may be modeled, and that numerical results thus obtained can agree reasonably well with estimates based on pure science. Furthermore, the estimate helped confirm a consistent decline of more than 90 percent from historical levels of abundance for almost all heavily fished marine species and ecosystems examined so far, a stunning finding with global implications. Similar studies are now rapidly expanding our historical reach for other species and fisheries around the world.

MacCall and Bolster, Alexander, and Leavenworth celebrate achievements, while Field and colleagues and Daniel Vickers emphasize the limitations of understanding in terms that are surprisingly similar. Biological and physical unknowns and unknowables, historical contingency, attenuation and lacunae, and the staggering entanglement of ecological and social processes confound the predictive ability of scientific models. Moreover, change takes place so rapidly today that the adaptive powers of individual species, ecosystems, and human cultures may now be outstripped.

Vickers's chapter presents a disquisition on historical thought, context, and uncertainty that is at once magisterial and profoundly personal. As a history professor at Memorial University, he saw the Newfoundland cod fishery collapse firsthand. His chapter begins with his own memories and reflections about that crisis and ends with deliberations on its political and cultural underpinnings. Like Bolster, Alexander, and Leavenworth, he distinguishes between the uses of the past based on accumulated evidence and the uses of history, which is the story of humanity ever filtered and interpreted through our own sense of ourselves. He cautions that cultural values are historical—they cannot be ignored with impunity in scientific models or in policymaking—and that people who are separated from their livelihood by historical contingency as much as by institutional folly and personal choice deserve compassion.

Chapter 6

The Historical Abundance of Cod on the Nova Scotian Shelf

W. Jeffrey Bolster, Karen E. Alexander, and William B. Leavenworth

Only a generation ago marine scientists, fishery managers, and maritime historians shared the popular assumption that diminished fish stocks and damaged marine ecosystems were lamentable artifacts of the late twentieth century, of synthetic filaments, fish finding sonar, and electronic navigational systems. It seemed highly unlikely that historic sailing fleets could have depleted naturally abundant fish populations with simple hooks, hemp line, and handmade nets.

Times have changed. Recently scientists, historians, journalists, and fishery managers have come to the realization that, while destructive anthropogenic impacts on the ocean have accelerated dramatically since World War II, this phenomenon has deep roots, and deeper implications. As governments and NGOs grapple with the extent of the fisheries crisis and attempt to implement policies that promote restoration of degraded marine ecosystems, questions arise about the extent of the damage and about appropriate targets for rebuilding fish stocks. Just what constitutes a healthy marine ecosystem or a healthy population of a desirable species like cod? What sort of baseline populations existed at certain points in the past? What is the magnitude of current problems? Questions like these can only be answered by incorporating historical perspective.

This chapter presents the results of collaboration by maritime historians and marine ecologists assessing one part of the nineteenth-century Northwest Atlantic cod fishery. It expands upon a paper previously published by our interdisciplinary group. Our subjects are New England fishermen in the age of sail and the cod they caught with hand lines on the Nova Scotian Shelf. Our group discovered that data recorded with quill pens in stained logbooks during the 1850s proved suitable for computerized stock assessment models and distribution analysis using GIS. The analysis relies on historians' assessment and contextualization of archival documents pertinent to the fishery, notably thousands of historical fishermen's logbooks that can be linked to other customs house records to provide fleet size, landings in weight and numbers, fishing effort, and geographic location.

Our results are noteworthy in several ways. We have calculated the earliest biomass estimate for a fished cod stock anywhere in the world using data from the 1850s for the Scotian Shelf. We have reconstructed the history of localized overfishing by the Scotian Shelf fleet during the 1850s, a pivotal decade during which New Englanders reluctantly adopted a new longline technology—dory tub trawling—in response to declining catch and overpowering competition by large French factory brigs. And we have developed an interdisciplinary methodology that may inform future collaborations.

In the process of this journey, members of our group adopted new terms from unfamiliar lexicons, came to appreciate very different modes of thinking, and learned to frame new questions in new ways. All agree that our results would have been impossible without genuine collaboration. This chapter reviews the literature that inspired the project, explains our textual and scientific analysis of the historical documents, and presents the results. We see it as validating the pioneers who first called for this type of work, secure in their conviction that knowledge of oceans past is critical to the future health of the sea.

Background

It is now clear that historical overfishing damaged coastal ecosystems and reduced fish stocks before scientists could measure its magnitude or observe its effects. Yet until recently ecological changes went unnoticed because no one was looking for them, even though overfishing was identified as an *economic* problem in Europe and the United States during the mid-nineteenth century. The first systematic fisheries studies in the United

States, at both state and federal levels, responded to existing public concern about decreasing catches, partially attributed to overfishing. Similarly in Europe, as Tim Smith has shown, dramatic fluctuations in Norway's Lofoten Islands cod fishery during the 1860s prompted a government inquiry and the foundation of fisheries science. Overfishing not only preceded scientific inquiry into fisheries productivity, but was a precondition to it.

Accurately assessing the changing nature of marine abundance has been difficult, however, because early narrative accounts lacked precision. Awed by the unfathomable quantities of fish he found in the Gulf of Maine on his first voyage there in 1614, Captain John Smith enjoined Englishmen to convert those seemingly inexhaustible stocks into profitable commodities. On the second edition of his map of New England in 1635, engravers added a vast school of fish between Cape Cod and Cape Ann, Massachusetts (figure 6.1). By 1675, there were reportedly 440 boats and about 1,300 men fishing the coast between Boston and eastern Maine, producing over 6 million pounds of dried salt cod annually. On the eve of the American Revolution, dried cod was the fourth most valuable export in the American mainland colonies, sent to Morocco, southern Europe, the wine islands of Madeira, the Azores, the Canaries, and the Caribbean. Incontrovertible evidence indicates that humans were affecting regional populations of right whales, great auks, sea minks (a North American mustelid, substantially larger than the American mink, extinct by the mid-nineteenth century), and various birds, and recent research indicates that early Puritan fisheries had perceptible impact on cod stocks south of Cape Ann. However, few written records remain from seventeenth- and eighteenth-century New England fisheries. We simply cannot quantify that era's abundance with the sleek precision of a number.

Following the American Civil War, New England fishermen worried that fish were growing scarce. But how scarce? Prominent politicians in Maine and Massachusetts called for better information about fish biology and for government regulation of fishing gear. They feared overfishing would harm commercial stocks and reduce the livelihood of coastal communities. In 1872, Maine's Fish Commissioner, Elias M. Stilwell, wrote to Spencer Baird, head of the new federal Fish Commission, asking for information on "the probable cause of the rapid diminution of the supply of food-fishes on the coast of New England." Six years later in his State of the State address, Governor Selden Connor of Maine warned that depleting "river fish," including anadromous species like alewives, could reduce valuable populations of pollock, cod, and haddock that entered coastal waters to feed. Yet along the Penobscot Bay estuary, a U.S. Fish Commission map

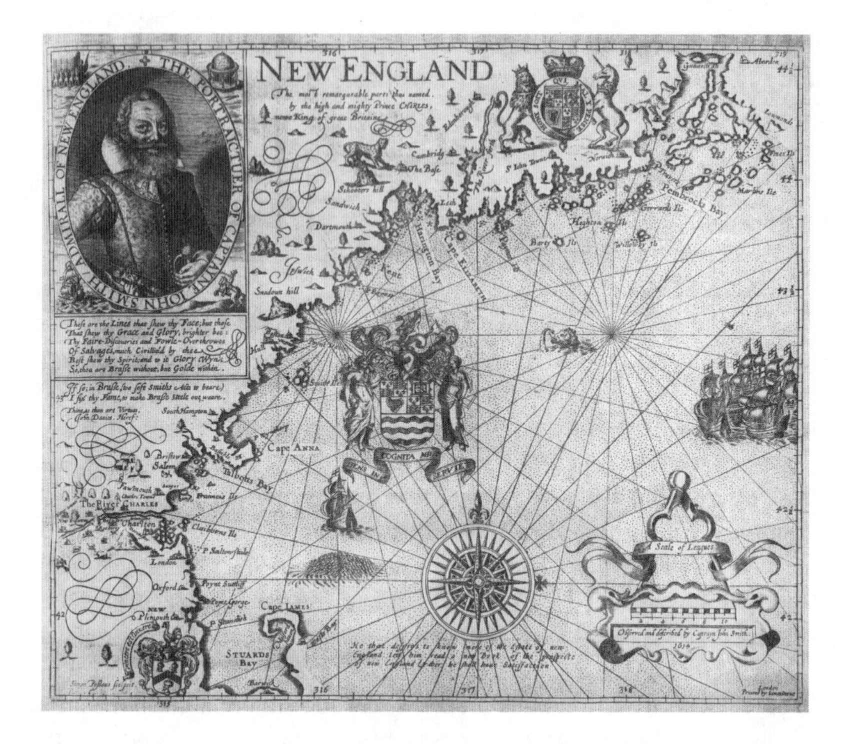

FIGURE 6.1. Map of New England, 1635. John Smith drew the region he explored and named in 1614, and first published his map in 1616 in a *Description of New England*. He praised the abundance of cod, whales, and other valuable marine species easily caught in coastal waters. This second edition of the map adds a great shoal of cod roiling up to the surface near Smith's vessel lying just east of Cape Ann and Cape Cod, near what today is the Stellwagen Bank National Marine Sanctuary.

located 44 pound nets and 150 weirs actively harvesting "river fish," as well as former sites already abandoned. The same map noted that the number of local fishworks had been reduced from 17 in 1832 to just 3 in 1873. Obviously, Spencer Baird's scientists in the 1870s inherited ecosystems already anthropogenically altered. In New England, home to the oldest fisheries in the United States, no baseline exists for a marine environment in a pristine state.

Hoping to chart the changing nature of abundance, the first generations of marine biologists in the United States and Europe began to collect landings data. Not until the late nineteenth century, however, were landings from the New England fleet regularly tallied. While this initiated valuable time series for the twentieth century, the coast of New England had been fished since about 1600. Few reliable catch statistics existed prior to

1900. Any that could be reconstructed from historical documents would lend an especially valuable perspective to the changing nature of fished stocks.

Early fishery scientists based their new discipline, in part, on fish stories. As late as the 1950s scientists like Henry Bigelow, instrumental in the development of American oceanography, were still listening to fishermen. It was commercial fishermen and saltwater anglers, noted Bigelow and his collaborator, William Schroeder, authors of the magisterial *Fishes of the Gulf of Maine*, who "supplied us with a vast amount of first-hand information on the habits, distribution, and abundance of commercial and game fishes, which could be had from no other source." By the late twentieth century, however, the disciplines of ecology and fisheries science had changed significantly: descriptions of systems or parts of systems (including basic taxonomy) were no longer particularly valued. Mathematical modeling became the preferred route to professional success. The evolution of the field thus tended to suppress descriptive natural histories. Few modelers, moreover, sought data indicative of long-term historical change because gaps in the data were messy and collection methods unreliable.

By the 1970s, when the first scientifically inspired federal regulations limited American fishermen's access to fish, commercial fishermen and scientists were at loggerheads. Scientists found it increasingly difficult to believe fishermen or to imagine that fishermen had ever been believable. The result of these changes was that by the late twentieth century, as fisheries were collapsing all over the world, marine ecologists had long since suspended attention to stories and anecdotal evidence about the former state of the seas, despite a desperate need for information about oceans past.

Then Daniel Pauly challenged fisheries scientists to think historically and to incorporate narrative evidence from the past into ocean ecosystem models and fisheries management plans. The "shifting baseline syndrome of fisheries" in his title ascribed deflated biological reference points and debased ecosystem standards to the human proclivity for assessing change in terms of personal experience or living memory. Most scientists, he suggested, failed to take long-term changes into account. This oversight led to a gradual, unwitting "accommodation" of species decline and ecosystem degradation in scientific models and management plans. As Pauly argued in 1995,

> Developing frameworks for incorporation of earlier knowledge—which is what the anecdotes are—into the present models of fisheries scientists would not only have the effect of adding history to a discipline that has suffered from lack of historical reflection, but also of

> bringing into biodiversity debates an extremely speciose group of vertebrates: the fishes, whose ecology and evolution are as strongly impacted by human activities as the denizens of the tropical and other rain forests that presently occupy center stage in such debates. Frameworks that maximize the use of fisheries history would help us to understand and to overcome—in part at least—the shifting baselines syndrome.

Jeremy Jackson and others subsequently revealed in more detail the perversion of standards by which the "natural" ocean had come to be evaluated. For centuries indigenous peoples had hooked, speared, trapped, netted, or poisoned marine mammals and large carnivorous fishes, and laboriously dredged up shellfish. Archaeological evidence has shown that a few marine species such as seals and sea otters in the Pacific Northwest were hunted so heavily that local coastal ecosystems were affected. In the European colonial period, which began in earnest during the sixteenth century, intensive commercial fishing and whaling with comparatively primitive equipment expanded worldwide in the wake of settlement, first in the Americas, then in Africa, and finally in the waters around Australia and New Zealand. Pandolfi and colleagues ultimately attributed the "collapse" of coastal and coral reef ecosystems to the impacts of overfishing and the concomitant loss of certain trophic levels. The implication was that "fishing down the food web" had been abetted by the shifting baseline syndrome.

As these authors and others made the case for marine ecologists to think historically, an influential group of historians was already challenging the historical profession for paying insufficient attention to environmental change. Led by scholars such as Alfred Crosby, Donald Worster, William Cronon, and Richard White in the United States, and Richard Hoffmann in Canada, this insurgency propelled the development of environmental history, making it one of the fastest growing subfields in American history after the 1980s. The lion's share of this work was terrestrial environmental history, but the challenge to historians to think ecologically was not lost on all maritime historians. Our interdisciplinary collaboration was nurtured by injunctions from colleagues that ecology become more historical and history more ecological.

The Historian's Perspective

At the heart of this volume are a couple of key questions: Why is the past important? Why is history important? They are good questions, but they are not identical. If we honor the past and use it to inform current environmental debates, we need to be clear about these terms. The past is gone. In its vastness and messiness it is largely unknowable, even though the present and future are contingent upon that past. We can never experience the past. It eludes our grasp as effectively as a fistful of water.

History is something different. Like ecology, it is a form of analysis. Most history is an interim report on some small slice of the past based on the best available evidence and on currently respected methodology, verifiable, yet subject to revision when better resources and methods become available. In that, it is like ecology. In its reliance on narratives, however, both as a means of conveying what is known and as sources to be used in the investigation, professional history differs considerably from professional ecology. Historians often, but not always, use narrative to tell what they know. Historians, moreover, frequently rely on previous narratives for information, sometimes in the form of stories left by the people they study. Historians try to link the experiences in those stories with the meanings imposed on them by the people the stories are about. The point is to honor historical actors on their own terms rather than recasting them as people "just like us."

The "anecdotes" in Pauly's 1995 title referred to the sorts of nuggets regarding the past on which some historians traditionally have relied. However, most scientists have commonly dismissed anecdotal evidence for three reasons. First, anecdotes relate to singular events and are difficult to incorporate into models. Second, no guarantees of accuracy exist. Finally, anecdotal information is often conveyed descriptively or comparatively rather than quantitatively. In a famous example, Milan's envoy in London, Raimondo di Socino, reported on John Cabot's return from his first trip to Newfoundland in 1497 and included one of the earliest accounts of the abundance of fish in the New World. "The Sea there is swarming with fish," noted the diplomat, "which can be taken not only with the net but in baskets let down with a stone." Such anecdotes convey magnitude by analogy, narrative, and context so that it becomes memorable, but arguments about accuracy are inevitable in the absence of quantitative measurement.

Of course historical sources consist of much more than anecdotes. Quantitative data derived from historical records can provide reliable statistics, but the raw data frequently require substantial interpretation before it is usable. The "new social history" that came of age during the 1960s unleashed what was then called the cliometric revolution, the merging of quantifiable analysis with Clio, the muse of history. Following E. H. Carr's dictum that history should "become more sociological"—by which he meant more quantitative—practitioners began to examine a host of records from the past: tax records, voter rolls, militia musters, crew lists, probate records, land transfers, slave manifests, and other documents that were subject to mathematical assessment. The point was not only to create statistical profiles, but to use materials from the past that had been generated by daily record keeping, thus avoiding the bias intrinsic to self-consciously created texts such as letters, diaries, or anecdotes. In a classic case, Nobel Prize–winning economist Robert Fogel and his colleague Stanley Engerman discounted historical context in their statistical analyses of the economics of chattel slavery, *Time on the Cross*. They concluded that slavery was economically viable, that most slaves were adequately treated, and that some benefited from their own labor. Controversy erupted in the historical community. Fifteen years later in a subsequent book on the same subject, *Without Consent or Contract*, the authors contextualized their analysis, correcting erroneous interpretations and solidifying sound statistical arguments. It became a milestone in economic history because context mattered.

No matter what sources are used, historical analysis always rests on incomplete documentary evidence that survived the random winnowing of fire, insects, disposal, and neglect. While historians search for "the smoking gun," they understand that they will never have all of the information they want. Good historical reconstruction is like paleontology, where much can be made from one bone—it is the best interpretation that can be made with the materials at hand.

The Reliability and Nature of Nineteenth-Century Cod Fishermen's Logbooks

Our project began with a fortuitous discovery by a University of New Hampshire master's student in History, Robert Gee. While researching a seminar paper in marine environmental history he found 233 nineteenth-century fishing logs in the James Duncan Phillips Library at the Peabody Essex Museum in Salem, Massachusetts. His perceptive appraisal of these logs' value led to the discovery of about 1,500 more in the National Archives Regional Administration facility at Waltham, Massachusetts, and in other public and private collections. This remarkable run of logs spans the years 1852 to 1866 and records tens of thousands of numerical entries pertaining to the abundance and distribution of cod in specific regions of the northwest Atlantic during those years. Almost all logs provide the daily catch of each fisherman and name of the bank on which the vessel was fishing. Sometimes navigational coordinates are indicated. Catch per day, with reference to place, is thus recoverable (figure 6.2).

Thousands of fishing agreements, licenses, bonds, enrollments, and other ancillary documents also exist that can be linked to maximize the available data. Time consuming, and sometimes logistically difficult, these linkages are essential. For instance, fishing agreements, the seasonal contracts between the vessel and the crew, remain for many of the trips for which logs exist. They provide the size of the fishing vessel in terms of tonnage and the number of quintals landed per fare. (A quintal consisted of 112 pounds of dried cod. Each separate trip in a season was called a fare.) Linking the two sets of records provides the tonnage of the vessel, the size of the crew, the daily catch per man, and the total weight of that catch—all of the essentials necessary to calculate catch per unit effort (CPUE) with reference to time and place for this mid-nineteenth-century cod fishery.

The logs are also rife with anecdotal information illuminating the nature of specific places in the sea. Some daily log entries give the depth and composition of the ocean floor. To experienced captains like Larkin West of the Beverly, Massachusetts, schooner TORPEDO, bottom conditions indicated the likelihood of finding fish. Skippers armed the end of their lead lines with a blob of tallow to retrieve bottom samples. His log for August 25, 1852, reads, "this night and morning caught 450, hove up, tried some time to find rough bottom." But a week later, "this night caught 140, the bottom being rocky and fish scarce." The log of the IODINE in 1856 recorded a number of distinct bottom conditions on Nova Scotia's Western Bank: "Lat obs. 43°55′, green sandy bottom (6/9); Lat 43°55′, 35 fath[oms], moss bottom (6/13); shift 2 mi., anchor 32 fath., pumpkiny bottom (6/24); here we find plenty of fish (6/25); anchor 33 fath., rough mussel bottom, fish very large, Lat obs. 43°55′ (6/28)," the last noted with satisfaction.

While the vast majority of the information in these logs concerns cod, other species are occasionally noted, especially species used for bait. When fishermen ran short of barreled salt clams shipped for the voyage, they set gill nets at night for herring or other baitfish. Seabirds, considered a harbinger of cod, were also shot for bait. The crew of the PETREL in May 1854, "got 163 herring. Got today 221 f[ish]." Two days later they had better fishing, "we got 535 large fish, use hagdons [seabirds, likely shearwaters] as bait." Calvin Foster, the captain and a true vocational scientist, recorded fishing success using different baits. He also occasionally carried a barometer, a rarity at the time. Other marketable species like hake or halibut sometimes made a welcome change from ships' provisions and cod chowder, but

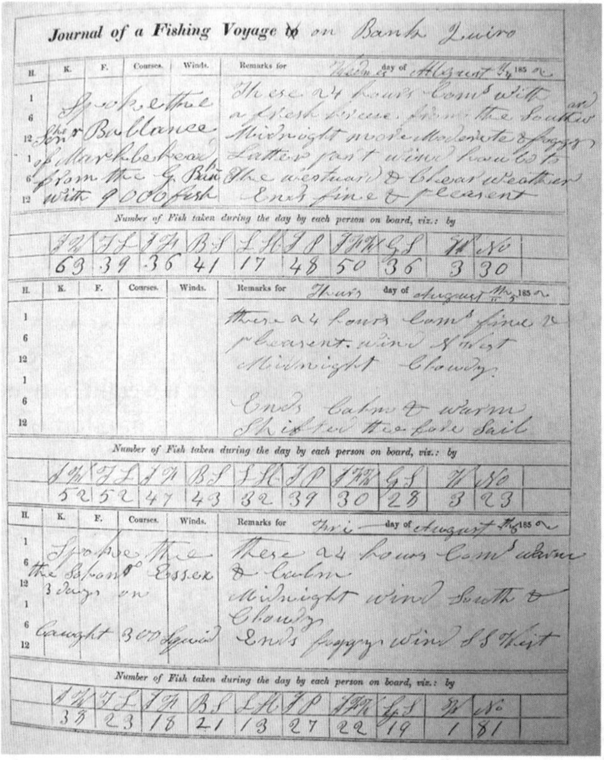

FIGURE 6.2. A page from the log of the Beverly, Massachusetts, fishing schooner DOVE in 1852, John Woodbury, captain. Most New England fishing captains recorded the events of one fishing season in a single, commercially printed logbook. This is the form most commonly used in the deepwater fishing fleets. Here the DOVE is anchored on Banquereau, or "Bank Quiro," August 4–6. Each fisherman's catch is recorded each day under his initials. Woodbury's remarks are concerned mostly with the weather, although he "shifted the foresail" on August 5. He also spoke with the schooners BALANCE of Marblehead, from the Grand Banks with 9,000 fish, and the ESSEX, likely of Beverly. On August 6, the log notes that the DOVE caught 300 squid, which would have been used for bait. (Courtesy of National Archives and Records Administration, Waltham, MA)

they were not "fish" in the parlance of fishermen: (HENRY 9/29/1855 on Brown's Bank) "tried for fish, but caught nothing but haddock." Fish were cod.

These records form the single best archival collection for a nineteenth-century fishery of which we know, but it is far from complete. Of seventeen

New England customs districts, eight provide 97 percent of the logs. Almost all the surviving logs are from Maine and Massachusetts vessels, although large fleets also sailed from Connecticut, Rhode Island, and New Hampshire. That the records exist at all is almost entirely due to a long-term government subsidy of the cod fishery and legislative requirements for bureaucratic oversight between 1852 and 1866.

In 1792 the U.S. Congress passed legislation authorizing an annual bounty to cod fishing vessels. To qualify, a vessel had to pursue only cod for 120 days or more during the fishing season. Cod was far and away the most valuable commodity exported from New England in the late colonial period. Moreover, Congress followed the British Parliament's assumption that the fishing fleet was "a nursery for seamen" for the navy. A federally subsidized bounty for cod fishing thus seemed politically justified, especially in light of the fleet's disastrous state following the American Revolution. From fishermen's perspective, the bounty was an important part of seasonal earnings, offsetting the duty on imported salt necessary to preserve fish. Each district's customs collector paid yearly remittances out of his gross receipts to every vessel over 5 tons fishing for cod, whether it fished inshore or on the deepwater banks.

In 1852 the law renewing the bounty was changed. The new authorization stipulated that to receive the subsidy each captain had to submit a seasonal logbook to the Collector of Customs in the vessel's home port. That legislation was a boon for future historians. Not only were the logs standardized, but many were retained. In 1866, however, Congress discontinued the bounty because representatives from the southern and western states were no longer interested in subsidizing the New England fishery. After 1866, captains kept logs less religiously. More pertinently, government offices or archives no longer retained a critical mass of fishery logbooks. While scattered logs remain from years before 1852 and after 1866, the systematic run of quantifiable data exists only from 1852 to 1866.

Little incentive existed to falsify entries because of an unintentional system of internal checks. To begin with, the government encouraged fishing. Fishermen were neither taxed on their catch or restricted by quotas. And the federal bounty was paid according to the size of the vessel, not by the number or weight of cod landed. The captain did not need to show government officials anything except that his crew had pursued cod for 120 days. However, each man was paid based on the number of fish that he landed, and it was in his interest to see that each fish was recorded. When paid on shares, the usual procedure was that each man received the same percentage of the net proceeds as the fraction of total fish he had caught.

Experienced men with a knack for hand-lining stood to earn considerably more than lazy or incompetent fishermen. Since the log recorded the number of fish caught daily and the crew was not paid until the end of the season, the log served as the payroll account. Each fisherman had a vested interest in its accuracy. Finally, the proceeds from the season were based on the weight of the catch, not on the numbers of fish. The fish were weighed at the end of each fare and sold to a merchant at the end of the season by total weight, expressed in quintals. Regulations and customary procedures thus provided an interlocking system of internal checks that discouraged chicanery.

The logs also provide accurate records of vessels' locations and record encounters with other vessels, establishing the minimum size of fishing fleets on different banks. Fishermen were not secretive, but social. Vessels fished in loose groups that constantly changed composition as some left and others arrived. Often, congregating vessels hailed from the same home port or contained family members who were otherwise dispersed around New England and the Canadian Maritimes. Visiting and exchanging information was the norm. On one pleasant Sunday, generally observed as a day of rest by American and Canadian fishermen, the SARAH on Middle Bank (9/12/1852) "had a visit from a part of the RICHMOND's crew and a part of the BRIDE's crew and a part of the ANGLER's crew and a part of the MECHANIC's crew, and a part of the GEORGE's crew."

Meetings and conversations were anticipated eagerly and news spread quickly. When the cook on the FRANKLIN died on October 29, 1852, his death at sea was duly recorded in the log of the E. W. FORREST although the two vessels passed at a distance. Because he had learned about the cook's illness a week earlier, Captain Woodbury of the FORREST noted that they "saw the FRANKLIN underweigh, and tried to get to us, but the tide run so strong she did not, her colors was at half mast, supposed his cook was dead (10/29/1852)." Sightings took on great importance when a vessel was lost, like the GREENLEAF in 1852, last reported in the logs of the ESSEX and the HENRY on August 23. Such accounts provided the owners and families with the last known information about the schooner and the men aboard her. Everyone shared the solemn obligation to tell the truth. An extended system of shared vocational knowledge and cultural values worked to reinforce cooperation among the fleet on what had traditionally functioned, at least from the fishermen's perspective, as a maritime common.

Log keepers routinely recorded the "vessels spoken" in daily entries when pleasantries or information were exchanged. Often these entries followed a formula. They provide the spoken vessel's name, home port, and a

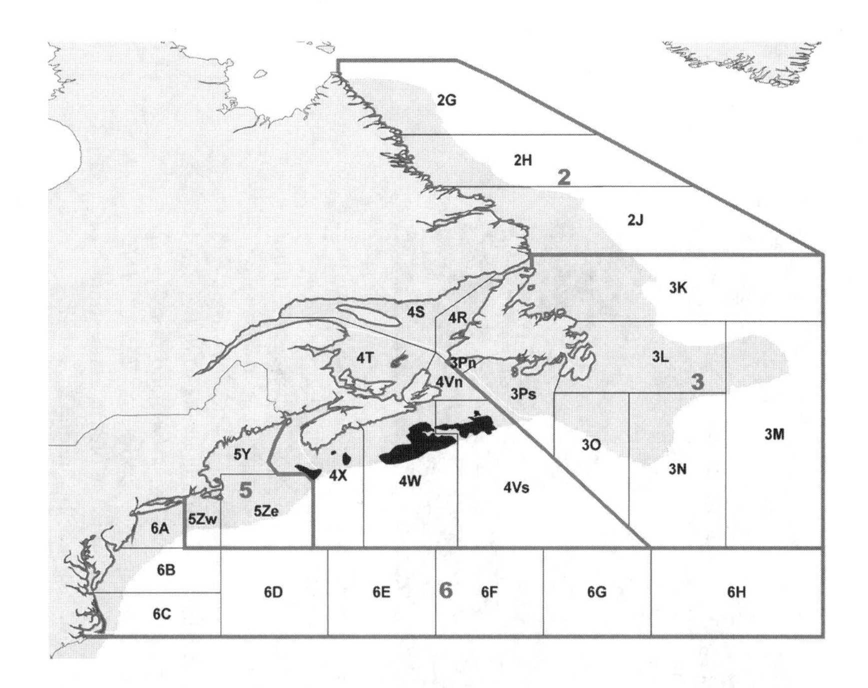

FIGURE 6.3. Global Information Systems (GIS) plot of large marine ecosystems (LMEs) [*shaded region delimited by thin white lines*], and North Atlantic Fisheries Organization (NAFO) zones [*outlined in gray, and numbered*]. LMEs, distinctive ecological regions extending from river drainage systems to the edge of the adjacent continental shelf, can cover 200,000 km^2 (Garcia and Farmer, editors [2000]). Large Marine Ecosystems. Garcia, S., and T. Farmer, editors [2000], LME #8: Scotian Shelf). NAFO zones are even larger. NAFO zone 4 encompasses the Scotian Shelf LME, but extends west to the U.S.–Canada border, and south to the approximate latitude of the Delaware River estuary (Halliday and Pinhorn [1990]). The dark gray regions in NAFO zones 4X, 4Vs, and 4W are the Scotian Shelf banks where the Beverly fleet fished in the 1850s. (GIS map by Stefan Claesson.)

Gulf of Maine, the offshore banks of the Scotian Shelf, the beach-based operations on the southern Labrador coast, the Bay of St. Lawrence, and the offshore Grand Banks. Comparatively few large vessels from the Beverly fleet in the 1850s spent the season on the Grand Banks.

Fishermen showed marked preferences for fishing familiar banks, with certain ports specializing in certain areas (figure 6.4). For instance, between 1852 and 1859, two-thirds of the Beverly vessels over 60 tons fished the Scotian Shelf full-time, primarily on two of its composite banks, Banquereau and the Western Banks. More than 90 percent of them fished there

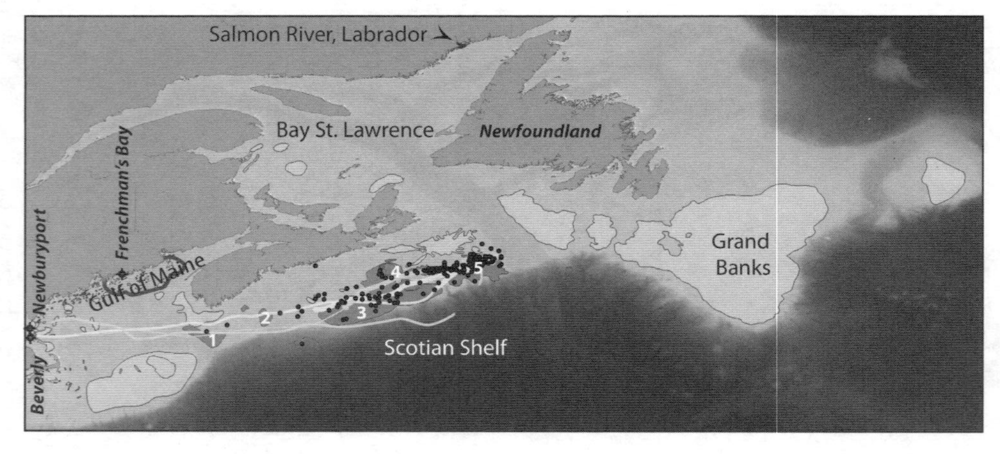

FIGURE 6.4. GIS map of fishing ports and fishing banks. Three New England ports with large collections of extant fishing logs. Beverly schooners fished the Scotian Shelf; large Newburyport schooners fished the Salmon River, Labrador; Frenchman's Bay vessels of all sizes fished inshore banks in the Gulf of Maine. Individual fishing grounds comprising the Scotian Shelf are (1) Brown's Bank, (2) Le Have and Roseway Banks, (3) Sable Island Bank, (4) Middle Bank, and (5) Banquereau. Brown's, Le Have and Roseway, Sable Island and Middle Banks were frequently lumped together and called the Western Banks. Course tracks to (*lighter line*) and from (*darker line*) the Scotian Shelf are plotted for the spring fare of the Beverly schooner ANGLER, in 1853, Nathan Buck, Jr., captain. Solid circles show where cod were landed. Note the concentration of effort on Banquereau. (Log courtesy of NARA Waltham. GIS map by Stefan Claesson.)

at least part time. In that same period, 60 percent of the vessels from Newburyport (most of which were quite large, over 90 tons burden) targeted one bay in southern Labrador, Salmon Bay (now St. Paul's River), near the Straits of Belle Isle. Small Newburyport vessels making up another 30 percent of that town's fleet cruised the Isles of Shoals and other inshore banks close to home. Likewise, from 1861 to 1865, 92 percent of the fishermen from Frenchman's Bay concentrated their efforts on the inshore fishery bounded essentially by Penobscot Bay to the west and Grand Manan on the New Brunswick border to the east. If captains could find fish on familiar grounds, they cruised there year after year. These strong territorial preferences suggested a connection between fishing patterns, home port, and fishing grounds, the kind of diagnostic correlation that offered promise as a statistical proxy.

After a preliminary study of each fleet, we decided to focus on the Beverly fleet fishing the Scotian Shelf from 1852 to 1859. There, we felt, the greatest possibility existed to reconstruct the precise geographic distribution of catch and fishing effort in the mid-nineteenth century. And with luck, it would be possible to calculate the size of the Scotian Shelf cod population in 1852.

Declining Catch and Technological Change

Neither historians nor biologists believed that primitive hook-and-line technology could affect the legendary abundance of species like cod, but the logbooks reveal they were wrong. During the early and middle years of the 1850s, 90–95 percent of the Beverly fleet's vessel-days per season were spent on the Scotian Shelf. But in 1859, many skippers shifted fishing to the Grand Banks or Gulf of St. Lawrence, voyages up to two times longer with substantially greater risk. Fleet size, however, remained roughly the same. The next year, many vessels withdrew from the fishery altogether. Between 1859 and 1861, the number of Beverly vessels fishing on the Scotian Shelf declined by half and the entire Beverly cod fleet by 55 percent. This was *before* the disruptions of the Civil War, during which the Massachusetts cod fleet dropped only 7 percent.

These dramatic changes in the deployment of fishing effort followed changes in catch per unit effort (CPUE). Catch per vessel for the thirty Beverly schooners in 1853 was down from the year before by about 1,200 fish. That would equate to slightly more than one-third of a typical fisherman's seasonal income. The fleet as a whole caught more fish, but the fleet was also bigger. The declines continued and, by 1856, the average catch per vessel had dropped 7,000 fish in just four years. The seven men fishing from an average schooner in 1855 caught what five men had expected to catch in 1852.

As catch declined, fishing patterns changed dramatically. From 1852 to 1854 the Beverly fleet concentrated efforts on Banquereau. Most vessels divided their season into two fares. Fishermen left Beverly on their spring fare in late April or early May, stopping briefly to fish on the Western Banks, but shifting to Banquereau by the third week in May. They filled their holds, or "wet their salt" as they put it, in about seventy-eight days, returning to Beverly by late July. After approximately two weeks in port unloading their catch, washing the vessel, undertaking routine maintenance, and fitting out with fresh stores, they sailed again in early to mid-August. Unlike the spring fare, vessels outward bound on the fall fare from 1852 to 1854 spent even less time on the Western Banks. Many proceeded directly to Banquereau. There they fished until mid- to late October, returning to Beverly in early November. At seventy-five days, the fares were roughly equal in length. In 1852 and 1853 there were two distinct peaks in fishing activity: one that lasted from late May until late June and another one that began in late July and lasted until late September. On a very good day on Banquereau in 1852 or 1853 a vessel like the CLARA M. PORTER could catch more than 1,000 cod.

By 1855 declining catches had affected the rhythm of the season and the distribution of vessels. A new pattern emerged that year. Beverly masters looked much more to the Western Banks to wet their salt, and they spent less time on Banquereau. During the first fare, the fleet divided its time equally between the Western Banks and Banquereau, leaving for Banquereau in early June, two to three weeks later than was common from 1852 to 1854. Numbers of vessels on the two banks were almost equal in both fares, and some vessels elected to remain on the Western Banks throughout. By the fall of 1857, more vessels fished on the Western Banks than on Banquereau (figure 6.4). Moreover, schooners averaged ten days longer each fare, incurring more risk.

The American economy was as much of a roller coaster in the nineteenth century as it is today, but even within that climate of volatility, 1857 was famous for its depression. We thought that the Panic of 1857 might have depressed fish prices. On the contrary, the price increased by about 25¢/quintal from the year before, and exports increased as well. Demand for cod remained high. However, since the average fisherman caught about 10 quintals fewer fish than he had in 1856, he brought home less money.

Although Samuel Wilson, aboard the LODI, still occasionally found "large schools of fish on top of water in every direction" (6/23 on the Western Banks), other skippers expressed concern. The ROBERT observed "good fishing weather, few fish on the ground. Would catch 7 or 8 good fish & all done. Dressed 102 fish, 40 or 50 good fish, the re[st] miserable devils" (6/8 on Banquereau). The SUSAN CENTER experimented early with tub trawls, but new gear did not help her find fish: (8/25 on Banquereau) "set trawls, found no fish. 30 sail of vessels in sight." (8/27) "fished in dories, found fish scarce." (8/31) "hove up for Middle Bank." (9/1) "Tried on Middle Bank, found no fish. Lat 44°30′."

Fisheries historians have long known that during the middle of the nineteenth century New England fishermen evolved their technique from simply hand-lining along the rail of their schooners, to hand-lining from small boats or dories, and then to long-lining (which they called tub trawling) from dories (figure 6.5). This story has generally been presented as a sequential tale of technological progress and a means by which fishermen could increase their catch. Little attention has been paid to the overlapping use of these technologies or to the ecological impetus for their adoption.

William Leavenworth has shown that transition to a significantly larger hook footprint, and then to a significantly larger number of hooks per man, was a critical turning point in the history of the fishery. It may have been more critical in the population history of codfish than has been recognized. This transition occurred during the 1850s, precisely at the time that catches were declining. During the transition from hand-lining over the rail of the vessel to tub trawling from dories, fishermen on individual vessels employed some or all of these methods during the same season, and often during the same day. Skilled hand-liners initially were skeptical of expensive and dangerous innovations. But all were aware that annual catches were declining.

The dominant technology in 1850 varied little from that of the seventeenth century, except that the schooners were larger, faster, and less forgiving. Hand-lining still prevailed. Leaving home with barrels of bait, usually salted clams, fishermen proceeded east in their schooners until they found the edge of the banks by sounding with a lead line. They tried for fish at intervals as they sailed or drifted across the banks until they found a place where fish were plentiful. Anchoring, each man fished with two or four hooks over the rail until the fish became scarce. They might stay a day or two, or more, until they weighed anchor and repeated the process. Using one or two double-hook hand lines fitted with size 12 or larger hooks, a skilled hand-liner adept at catching substantial numbers of large fish was an artisan with status by virtue of expertise (figure 6.5A, top).

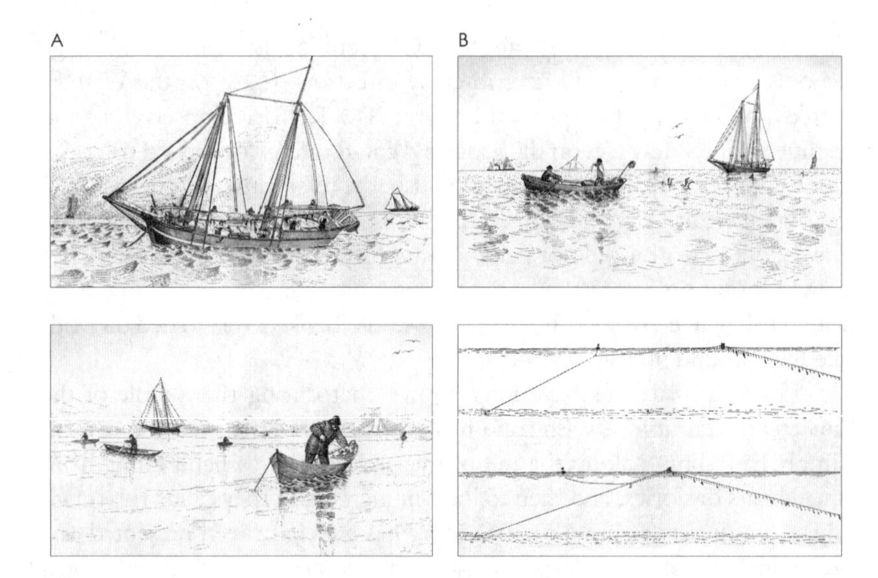

FIGURE 6.5. (A) Hand-lining. *Top*, Four fishermen are hand-lining over the rail of a typical small schooner. Hooks covered an area about the size of the vessel's hull. This technology was common on the Scotian Shelf before 1855. *Bottom*, One fisherman hand-lines from a dory. Other dories are nearby while their schooner rides at anchor somewhere in the distance. Adopting dory hand-lining greatly increased the hook footprint, the area over which one schooner could deploy its fishermen and hooks. (plates 23 and 24, section V in Goode [1884–1887]).
(B) Tub trawling. *Top*, Two fishermen in a dory set a trawl for cod. A typical dory set over 1500ft of ground line with 300 baited hooks. *Bottom*, The trawl was anchored to the bottom and attached to buoys for easy retrieval. Not only did tub trawling increase the hook footprint, it put more hooks in the water (Leavenworth [2006]; plates 26 and 27, section V in Goode [1884–1887]).

As cod became scarcer during the mid-1850s, particularly on Banquereau, some skippers adopted the practice of fishing from the stern boat, as well as from the rail of the schooner. This extended the area they could cover from a single anchorage. Initially, skippers sent their stern boat to look for fish, following with the schooner if the mission was successful. No extra outlay for equipment was necessary, but success at this level encouraged investment in dories. Some fishermen were increasingly interested in fishing from small boats, as the log of the RICHMOND from 1854 on Banquereau indicates: (6/15) "Sent the boat out. boat returned with 131 fish. We got altogether 520f." (6/16) "This day we got 300, we got 49 fish in the boat." In 1856 Captain Samuel Wilson expressed greater enthusiasm for boats: (7/8, Banquereau) "this day we catch about all of our fish in boats; they will not bite aboard the vessel." (7/12) "our boats do very well by going a long way after them."

Satisfactory landings from the stern boat encouraged fishermen to begin carrying several boats, eventually dories, on deck to the fishing grounds. While the immediate effects of this shift in technology were to increase their hook "footprint" at any given anchorage and to catch more fish, it also made fishing more dangerous. Schooners remained at anchor while much of the crew rowed away to find fish (figure 6.5A, bottom). Dories were easily lost, and schooners were threatened, too. Inability to weigh anchor rapidly in the face of a sudden gale jeopardized shorthanded schooners. Insurance policies reflected this concern in 1856 by stipulating that the schooner never be left untended. Nevertheless, the industry evolved from hand-lining along the rail to hand-lining from small boats. In 1853 fewer than 5 percent of the Beverly vessels fishing on the Scotian Shelf mentioned fishing from boats or dories. By 1857, 35 percent of Beverly vessels were using this new technology.

In 1858, as catches continued to decline, and as part of the New England fleet experimented with dory hand-lining, French trawlers appeared on the Scotian Shelf. Substantial brigs and ships, each carrying two large boats and tub trawls (longline technology then known as "bultows") with a total of about 4,000 hooks, these brigs and ships were the nineteenth-century equivalent of factory ships today. The comparison is valid in light of the fact that the typical Beverly hand-lining schooner was wetting only 14 to 28 hooks. French vessels were officially limited by treaty to fishing within 100 miles of the tiny islands of St. Pierre and Michelon, French territories south of Newfoundland. Their bultows had already raised the ire of Americans. Lorenzo Sabine, a congressional expert on the New England cod fishery, condemned bultows in 1853 and accused the well-financed and heavily subsidized French fleet of treaty violations by employing them. By 1858, however, France and Great Britain were in the midst of diplomatic negotiations regarding the fishery, and the French were optimistic that restrictions would be loosened in Britain's Canadian waters. Perhaps anticipating a successful conclusion to the negotiations, French trawlers crossed the Laurentian Channel to Banquereau in 1858. The effect was devastating on the already beleaguered Beverly fleet.

FRANKLIN (7/8/1858) "saw 20 sail of French ships in sight, 50 sail of sch[ooners]."

LODI (7/20/1858) "fish very scarce today. The French bothers us very much. they run their trawl all around us so they get most all the fish."

PELICAN (7/22/1858) Lat 44°42′, "Not much chance for fishing on account of so many trawlers." catch 282f. (7/23) Lat 44°44′, "There is not much chance for fishing this season, there is so many Frenchmen trawlers." (7/24) Lat 44°43′, "fine for fishing, the chance is small, being surrounded by French trawls." (7/25) "fine for the French trawlers." . . . (7/27) "Fine for fishing. French ships & brigs & schooners aplenty."

PRIZE BANNER (9/6/1858) "Boarded the French ship CHARLOTTE, w/160 thousand fish."

In 1852, the best year for fishing in these records, the average Beverly vessel brought home 26,000 fish. For a seasoned Beverly skipper to board a French ship that had already landed 160,000 fish must have been dispiriting, especially given that twenty such ships had already been sighted on Banquereau that summer. And their spider web of lines, thousands of yards long with hooks every six feet, interfered with New Englanders' hand-lining.

Anchored at each end, tub trawls were set between two marker buoys (figure 6.5B). With a tub trawl, two men and a dory could tend hundreds or even thousands of hooks, instead of just four. Setting and retrieving trawls required minimal fishing skill of the sort hand-liners respected, but it called for adept boat handling. In bad weather fishermen in small boats were unable to retrieve their trawls. Trawls were abandoned as the wind rose, and dory men had all they could do to avoid foundering. Since vessels generally fished in fleets, and staying at anchor was the only way to save expensive trawls, gales might cause closely anchored vessels to drag anchor and collide, with catastrophic results. Tub trawling clearly made the fisheries more dangerous than ever for men on the banks.

Tub trawling also had the potential to remove considerably more fish from the ecosystem. Not only could average landings per man rise, but the demand for fresh bait escalated as well. Collateral bait fisheries expanded, especially in Nova Scotia. It appears, moreover, that the cod population on the Scotian Shelf had been declining before the arrival of French trawlers. CPUE for the Beverly Scotian Shelf fleet had already fallen 37 percent in just six years, when trawling was uncommon. Although biology or climate may have contributed to the decline, hand-lining from schooners and dories appears to have made an impact on cod stocks.

Calvin Foster, the experimentally minded Beverly skipper who carried a barometer to sea in 1854 and a telescope in 1858, assessed the circumstances he faced like this: (PETREL, 7/15 on Banquereau) "A French ship anchored very near us, pick up part of a Frenchman's trawl full of stinking fish." (8/28) "Got 222 fish on the trawl." catch 390f. (9/7) "fish scarce. Hove up and run SSW 12 miles. Boarded the French Barque 'Charlotte' and bartered for 10 lines of trawl and 800 hooks. Latter part anchored on rough bottom." Among the first to use trawls in the Beverly fleet, Foster bought the gear from French competitors. But that year, despite his new tub trawls, he brought home only 11,000 fish, less than half his take seven years before. MAYFLOWER fared worse. After seventy-four days on the Western Banks and Banquereau in the summer of 1858 she brought home a pitiful 1,846 fish, a mere 57 quintals. EXCHANGE's log keeper lamented, "always no fish."

Even so, Beverly captains did not rush to adopt tub trawls as a progressive solution to their dilemma. In 1858 only 7.6 percent of the Scotian Shelf bankers mentioned trawls in daily comments, although 41.5 percent of them fished in boats. French fishermen at St. Pierre–Michelon had been using trawls since the 1830s. New England outer banks fishermen knew about them, and a few innovators like Calvin Foster and Samuel Wilson had experimented with them for almost a decade. Yet reluctance to abandon older methods generally remained the norm despite the fact that some skippers were sufficiently flexible to fish alternately with hand lines, dories, or trawls, or even to fish adrift now and then, and despite plummeting catches. In Swampscott, a small village near Beverly with an active inshore hand-line fishery, some experienced men agitated against the use of trawls. They complained that trawls gave unfair advantage to large, well-capitalized schooners, which caught too great a share of available fish. Throughout the 1850s Swampscott fishermen petitioned the Massachusetts legislature to outlaw trawls because they feared that haddock, cod, and other demersal species would soon become "scarce as salmon."

By 1859 Beverly vessels in the Scotian Shelf fishery were in desperate straits, landing less than half as many fish as they had in 1852. The average number of cod caught per vessel per season dropped from 26,217 to 14,414. Since vessels were 10 percent larger, efficiency had fallen off more than 52 percent. That year, the Beverly Scotian Shelf fleet decreased by more than half because captains accustomed to fishing there decided to try their luck elsewhere. Among those who persisted on the familiar Scotian Shelf grounds, 74 percent fished one long fare lasting almost five months rather than returning home in the middle of the summer, as had been the norm. Logs expressed the general refrain: "no fish."

The gloomy 1859 season, the worst that decade for the Beverly Scotian Shelf fleet, was notable in another way. Much of the fleet turned to tub trawls. After 1859, no deepwater crew hand-lined from the vessel exclusively. All had made the transition to using dories, and most had adopted the trawls that had been eschewed just a few years earlier. Anthropologists have identified distinct adaptations modern fishers make to declining catch. Changing their fishing patterns is typical, including variations in grounds, adoption of new gear, and reallocation of time. Analyzing historic fishing patterns in terms of days at sea and geographic distribution shows that Beverly skippers reacted to declining CPUE on the Scotian Shelf as modern fishers react when stocks become overfished.

By 1861, when Scotian Shelf catches had been decreasing for about a decade, those once-plentiful grounds were virtually abandoned. As economic historian Harold Innis explained years ago, the Assembly of Nova Scotia in 1861 laid blame for the collapse of the cod fishery on Banquereau "a few years since . . . [to] set line fishing [tub trawling], first practiced on it by the French and latterly by United States fishermen." Twenty-two years later, when George Brown Goode published his monumental seven-volume series *Fisheries and Fishery Industries of the United States*, he noted about Banquereau: "Not much fished at present by Americans." Yet a generation earlier it had been the destination of choice for most of the Beverly fleet and for hundreds of schooners from Marblehead, Barnstable, Portland, Portsmouth, and other ports. Goode's observation suggests that hook-and-line fishing techniques in the age of sail had the ability to impact cod abundance and distribution. More ominously, it may convey a New England fishermen's consensus in the 1880s that twenty years of light fishing had not brought catch up to acceptable levels. By then, contemporary writers like J. S. Collins observed that New England fisheries had lapsed into "decadence" as greater opportunities enticed young men away from fishing. The question requires further historical research.

Calculating the Biomass of Cod in 1852

The steady drop in CPUE for the Beverly Scotian Shelf cod fishery over eight years mirrored the log keepers' narratives of struggle and adaptation induced by resource depletion. Fishermen worked increasingly hard to catch fewer fish, and fishing pressure affected the cod population. A fishing technology assumed benign proved, at least temporarily and locally, to be unsustainable in a population considered abundant by modern standards.

The same drop in CPUE presented the potential to answer an important biological question: how many cod gathered on the Scotian Shelf in the mid-nineteenth century during the fishing season? A population estimate derived from these data would be the earliest population estimate for a commercially harvested fish, predating modern tables of catch statistics by almost a hundred years, and the earliest regular scientific sampling surveys by thirty. Such a baseline could fundamentally change perceptions regarding the nature of a North Atlantic shelf marine ecosystem before mechanized exploitation and significantly redefine targets for a rebuilt fishery.

Our group employed a modification of the standard regression model developed by D. B. DeLury in 1947 for estimating initial abundance of fish stocks based on total removals when cumulative effort is known. The Chapman-Delury method, developed in 1972 to assess populations of sei whales in the Antarctic, includes natural mortality in its calculation, which had not been considered in DeLury's original model. By adapting this well-established stock assessment model to our detailed historical data, we estimated the biomass of the cod population in 1852, the first year of the time series.

The Chapman-DeLury model scales abundance to total removals from the harvested population (figure 6.6). Of course, the 236 Beverly schooners fishing there full-time were only a small portion of the entire Scotian Shelf fleet. Ninety Beverly schooners fished across the Western Banks or Banquereau on their way to the Grand Banks or the Gulf of St. Lawrence. More important, 1,313 vessels from 76 other ports in New England and Canada appeared in Beverly skippers' daily comments, the spoken vessel fleet. (Although Beverly vessels also spoke each other frequently, for the purposes of this model, the spoken vessels will be defined as non-Beverly vessels without extant logs.) Total catch taken from the Scotian Shelf each year is the sum of the catch of these three groups:

1. the total catch from each Beverly vessel that spent the entire season on the Shelf;
2. a portion of the catch from each Beverly vessel that spent part of the season on the Shelf; and
3. some unknown amount of catch from each vessel spoken on the Scotian Shelf.

Obtaining the catch for the first two categories was straightforward. Logs give seasonal catch in numbers of fish, and the corresponding fishing agreements provide the total weight in quintals. For 236 Beverly vessels fishing the Scotian Shelf full-time, both documents survive and provide catch in weight and numbers. From them we calculated a weight-to-numbers ratio for cod caught there exclusively. The average number of pounds per fish was estimated separately for each year internally in the model so that uncertainty was propagated through into the total biomass estimate. While no fishing agreements exist for six full-time Scotian Shelf

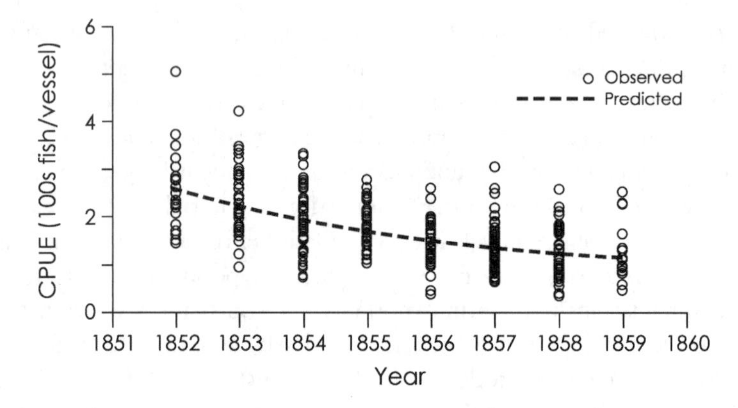

FIGURE 6.6. Chapman-DeLury graph. Data (*open circle*) and predictions (*broken line*) of catch per unit effort (CPUE) for Beverly fishing schooners on the Scotian Shelf, 1852–1859. CPUE is in (hundreds of fish)/(ton of vessel). (Rosenberg et al. [2005].)

logs, we estimated total catch weight for those vessels from their catch in numbers using the appropriate conversion. In logs from thirty-four full-time Scotian Shelf vessels, some daily catch figures were illegible and total catch impossible to calculate. Therefore, total catch in numbers was estimated from weight in quintals using the appropriate inverse ratio.

For Beverly vessels that fished across en route to distant grounds, our model assumed that only 25 percent of their catch in numbers of fish came from the Scotian Shelf. Logs show that these vessels spent, on average, more than 60 percent of their time on the Shelf. Since captains did not remain long on barren banks or grounds, future studies of the Grand Banks and Gulf of St. Lawrence fisheries should show that Beverly vessels caught a significant portion of cod outward bound and inward bound across the Scotian Shelf. Therefore, we reasoned, a conservative estimate for their Scotian Shelf catch would result in a conservative biomass estimate. We tested the sensitivity of our results to this assumption.

Estimating the total cod caught by the vessels spoken to for all the years in question proved more challenging. We knew the spoken vessels caught fish on the Scotian Shelf because many daily log entries preserved their cumulative catch up to that date. But we did not know how many cod they caught in total or what proportion were Scotian Shelf cod and relevant to our study. Their fishing patterns became an important question, whether they fished the Scotian Shelf full-time like two-thirds of Beverly schooners or passed through on their way to distant banks like the Labrador fleet of Newburyport. Modeling cod biomass depended upon knowing the size of spoken vessels (an indication of their fishing capacity) and determining their fishing profiles (decisions made at sea by captains based on experience, vessel size and capability, market, and the expectation of finding fish). Historical analysis of fishermen's behavior became a necessary component of the biomass model. Diagnostic characteristics of fishing profiles from different areas had to be established. Comparison with the profile of the Beverly fleet would determine if Beverly vessels could be considered representative of the Scotian Shelf vessels spoken and used as statistical proxies for the entire fleet.

We had already established a profile for all Beverly vessels relating tonnage to fishing patterns and captain's experience. That profile indicated where Beverly schooners and boats were likely to fish, for how long, and with what gear. Initial surveys of the 233 logs from Newburyport, Massachusetts, established fishing profiles for that fleet as well. For these two ports, vessel size essentially determined fishermen's technological and geographic options. The five Beverly vessels over 20 tons fishing in the inshore Gulf of Maine averaged 37.4 tons and their CPUE averaged 5.7 quintals/ton of vessel. Small schooners, sloops, or boats in the Gulf of Maine fleet usually made several short trips of a few weeks' duration during the season, coinciding with the expected arrival of migrating cod. They fished tiny banks just a few acres in area. Small crews needed little in the way of supplies, which could be replenished frequently at home or in nearby ports. Capital expenditure was minimal, and captains sometimes owned their vessels outright. Because they were more vulnerable to foul weather, few vessels under 50 tons fished offshore as far as Brown's Bank, the westernmost component of the Scotian Shelf.

Beverly vessels targeting the more distant Grand Banks were at the other end of the spectrum. They averaged almost 88.1 tons, and CPUE was 9.0 quintals/ton of vessel. Although they sometimes went into Newfoundland ports for bait or provisions, they generally carried supplies to last for months, spending nearly five months at sea in one long fare. Capital expenditure was commensurately greater. Owners and managers operated multiple vessels, and investment patterns encouraged consolidation. The shift to tub trawls happened in large offshore vessels first, the vessels requiring more manpower, a bigger crew, and much greater capital investment.

In Newburyport, tonnage also correlated with fishing ground. Very large schooners and brigs (more than 90 tons) voyaged to the Salmon River, Labrador, generally catching more than 100,000 very small cod each in a stop-seine bay fishery (figure 6.4). Small schooners, sloops, and boats (less than 45 tons) fished close to home in the Gulf of Maine.

A consistent pattern thus appeared in the Massachusetts fleet: vessels of the same size from the same community favored the same fishing grounds. Eighty-eight percent of all Beverly vessels and 90 percent of Newburyport vessels exhibited an identifiable fishing profile based on tonnage and home port. Individual Massachusetts captains likewise showed a marked preference for specific grounds.

The Frenchman's Bay, Maine, logs from the 1860s revealed a different profile, however. Although the average size of vessels fishing locally declined for the period, both large and small vessels fished inshore (figure 6.4). Among all of the vessels spoken by Beverly skippers, only 19 percent came from Maine. This percentage declined steadily from a high of 24 percent in 1852 to 8 percent in 1859. From 1860 to 1866, only 39 vessels from the twelve towns in the Frenchman's Bay Customs District fished offshore banks at all. Although the tendency to fish local grounds may have been exacerbated by the Civil War, the low number of Maine schooners spoken on the Scotian Shelf is not an aberration. Only 12 percent of spoken vessels hailed from Connecticut, Rhode Island, and Nova Scotia. Most of the Scotian Shelf fleet consisted of schooners based in Massachusetts.

Having determined the relevance of tonnage to fishing profile for vessels with extant logs, we turned to the 1,313 vessels spoken for which tonnage was the only statistical measurement obtainable. Using fishing agreements, licensing records, local newspapers, and other ancillary sources, we found tonnage measurements for 571 schooners in the vessels' spoken fleet—43 percent of all spoken vessels.

Massachusetts ports on Cape Ann and Cape Cod contributed 69 percent of the spoken vessels. These two distinctive peninsulas offer poor soil for farming and boast long fishing traditions. Fishing agreements listed tonnage and total catch weight for the Marblehead and the Barnstable Customs Districts, the largest home port cohorts next to Beverly. These agreements enabled us to profile the tonnage of 478 Massachusetts vessels spoken during this period. In 1852, Maine Scotian Shelf vessels were approximately the same size as the Beverly vessels, but their tonnage decreased as the decade progressed. Because Maine vessels showed up infrequently and declined steadily in numbers and size, they contributed little to the profile of the fleet.

Although the success of individual vessels depended to some extent on skill and chance, historical context indicates strongly that vessels of similar tonnage fishing the same grounds were likely to use similar gear and similar bait, and to exert similar fishing effort. Table 6.1 arrays the average tonnages of the fleet. It shows the average tonnage of Beverly vessels fishing full-time on the Scotian Shelf to be 2.1 tons less than that of the spoken vessels, a negligible amount. The Beverly fleet is thus representative of all the vessels spoken, irrespective of home port. To calculate biomass we assumed that the Beverly catch profile could be used as a proxy for that of the vessels spoken. The probable catch removed by the whole fleet became a function of the Beverly catch.

Adding the time spent by the 236 Beverly vessels that fished the Scotian Shelf full-time and the 90 Beverly vessels that fished it part-time (while passing through en route to the Grand Banks or the Gulf of St. Lawrence), we determined that 72.3 percent of the fleet's fishing days were spent on the Scotian Shelf between 1852 and 1859. We rounded up and assumed that 75 percent of the days fished by the vessels spoken were spent on the Scotian Shelf—or 75 percent of the spoken vessels fished full-time there—and caught the same amount of fish Beverly full-time vessels caught. While catch from Beverly vessels fishing across the bank was included in the model, catch from the vessels spoken assumed to be passing through was not. This made our assumptions conservative. Since Beverly "part-time" vessels still spent 60 percent of their time there, "part-time" spoken vessels may have caught more fish than we estimated. This is important because it relates to scaling the biomass estimate. If total catch was greater, then the biomass of cod in 1852 must have been greater as well.

For this biomass calculation, standard biological assumptions included a cod population in an approximately unfished state at the beginning of the time series, in 1852. Although Europeans had exploited those grounds since at least 1539, we assumed a minimal impact for this preindustrial fishery, much like the earliest stages of modern exploitation of an unfished species. The cod population is understood to be in equilibrium, with juvenile cod attaining breeding age as fast as adult cod die from natural causes. Over short time periods the annual instantaneous rate of recruitment and the annual instantaneous rate of natural mortality could be considered constant for an unfished species at an early stage of exploitation. We assumed a standard mortality rate of 0.2, the standard rate assumed for cod in modern stock assessments, but explored the sensitivity of our results to this assumption by calculating biomass for a range of mortality rates. Additionally, we examined the sensitivity of our results to the assumption that the recruitment rate equals the natural mortality rate by varying that proportion. With these conditions implicit, average population size for a given year became a function of that year's catch.

We related the decline in CPUE to the cumulative catch of the whole fleet for each year between 1852 and 1859 inclusive, and adjusted for assumed rates of natural mortality (M) and recruitment. We fit the

TABLE 6.1. Average vessel tonnage each year with standard deviation [sd].

Year	*All vessels spoken*	*sd*	*Massa-chusetts vessels*	*sd*	*Maine vessels*	*sd*	*Beverly Scotian Shelf fleet*	*sd*
1852	75.58	14.88	75.05	14.08	69.36	10.86	72.87	12.44
1853	79.54	15.41	80.16	14.34	77.42	24.40	76.47	15.41
1854	79.43	15.62	80.87	18.27	67.50	16.18	76.19	14.04
1855	83.57	13.83	85.07	14.03	75.07	25.38	78.33	14.13
1856	83.34	16.32	85.71	14.22	63.85	21.04	78.57	13.15
1857	84.75	14.31	85.77	13.86	70.00	18.75	81.32	15.47
1858	83.22	15.63	83.45	14.35	80.83	16.57	80.95	15.81
1859	83.71	15.85	86.98	13.83	62.60	9.12	81.78	17.21
Average	81.64	15.23	82.88	14.62	70.83	17.79	78.31	14.71

population dynamics model to an index of abundance I_j with the following equation:

$$I_j - kN_0 - k\left(\sum_{i=1}^{j-1} C_i(1-M)^{j-i} - \frac{1}{2}C_j\right)$$

where N_0 is the initial abundance, C_i is the catch in year i, M is the instantaneous natural mortality rate, 0.2, and k is constant of proportionality.

For the index of abundance, we used CPUE only from those Beverly vessels that spent the entire season on the Scotian Shelf and whose logs gave catch in numbers of fish. CPUE equals the number of cod caught in one fare divided by the number of days spent fishing multiplied by the tonnage of the vessel (catch/[days*tons]). The parameters of this model, including the annual average number of pounds per fish, were estimated by robust regression techniques contained in the AUTODIFF language incorporated into AD Model Builder. By estimating the parameters of the model, we solved for initial abundance, N_0, the number of cod on the Scotian Shelf in 1852. Total initial biomass was calculated as N_0 times the estimated average weight of a cod caught in 1852.

Quintals measured split, salted, and dried fish. Live biomass was derived using a modern conversion factor of 4.9 processed to live weight, which was calculated in the 1950s by the Canadian government. There is uncertainty in this conversion factor. Since 2005, other values converting dried to live weight have been found in historical literature as far back as the 1700s. Variations may represent regional differences, the size and oil content of the fish, curing methods, and even time at sea, and we are integrating this information into our current work. While changing the conversion factor changes the biomass estimate proportionally, abundance in numbers of cod remains the same.

The Biomass Estimate and Its Significance

Taking the sensitivity of our assumptions about mortality, recruitment, and total catch into account, and based only on New England fishery records, we estimated a biomass of 1,260,000 metric tons of cod on the Scotian Shelf in 1852. With respect to the Beverly part-time vessels, we found that ignoring their contribution to total catch only lowered the estimate by 4,000 metric tons. Including them with the full-timers only raised the estimate by 7,000 metric tons. Increasing the estimated instantaneous natural mortality rate produced less than proportionally lower estimates for the 1852 biomass. For example, increasing M by 50 percent only decreased the 1852 biomass by 25 percent to 947,000 metric tons. Similarly, choosing a smaller value for M produced less than proportionally larger estimates for the 1852 biomass. By assuming the reproductive rate was greater than the instantaneous natural mortality rate, we got lower estimates for the 1852 biomass, and vice versa, though the relationship was not linear. Assuming the reproductive rate equalled 125 percent of M decreased our biomass estimate by 26 percent, but an assumed rate of 75 percent for M increased cod biomass by 54 percent.

Not surprisingly, the biomass estimate proved very sensitive to the size of the vessels spoken fleet because total fleet size directly affects total catch. This encouraged us to craft our assumptions to be both reasonable and conservative. Data from the mid-nineteenth-century Nova Scotian inshore fisheries were not included in our estimate because we know of no comparable historical records from that fishery. In 2006, we obtained French removals from 1858 and 1859. A biomass estimate including catch from these fisheries might well have been much higher.

Myers and colleagues have estimated the hypothetical carrying capacity of the Scotian Shelf for cod based on productivity data. The western

Scotian Shelf—what historic Beverly fishermen called the Western Banks—has a carrying capacity of 232,715 metric tons of cod. The eastern Scotian Shelf—referred to as Banquereau in the 1850s, when it was heavily fished—has a carrying capacity of 917,789 metric tons of cod. The makes the total carrying capacity of the Scotian Shelf 1.15 million metric tons. This ecological estimate using entirely different methodology lies within the confidence interval of our biomass estimate derived from historical sources (figure 6.7). Moreover, carrying capacity estimates were based on data collected since 1950. The banks have been fished since at least 1539; therefore, it is likely that their productivity was even greater in the distant past.

Canada's Department of Fisheries and Oceans (DFO) has estimated total cod biomass in NAFO divisions 4X, 4Vs, and 4W since 1970. Fishing pressure from foreign factory ships was reduced with the imposition of the 200-mile limit in 1976, and peak total biomass in the era for which statistics are available occurred during the 1980s, at approximately 300,000 metric tons. This area includes not only the Scotian Shelf but also the inshore banks near Nova Scotia, the Bay of Fundy, and waters as far south as Cape May, New Jersey. The 2002 DFO estimate of total cod biomass for this area is less than 50,000 metric tons—a mere 4 percent of the biomass in 1852 (figure 6.7).

Our biomass estimate likely reflects only the adult cod population. Beverly fishermen in the 1850s complained when they caught small cod, and even shifted berth to avoid it, because small cod were worth less. The large hooks deployed in this deepwater fishery, sizes 10 and 12, reduced the landing of juveniles. The average weight of a salted, split, and dried fish was 4 pounds, meaning that live fish averaged about 20 pounds each. It is reasonable to assume that the fishermen we studied caught only adult cod and that our biomass estimate does not include juveniles. The DFO's 2002 estimate for adult cod biomass in NAFO divisions 4X, 4W, and 4Vs was only about 3,000 metric tons, less than 0.3 percent of the biomass of adult cod in 1852—a difference of three orders of magnitude (figure 6.7)!

A further comparison draws an especially stark picture. Records reveal that 43 Beverly schooners fishing the Scotian Shelf in 1855, just 20 percent of the vessels fishing there that year, wetting fewer than 1,200 hooks, caught 7,800 metric tons of cod. This is 600 metric tons more than the landings of the entire Canadian fleet in 1999. In fact, the entire *adult* cod biomass on the Scotian Shelf today, as estimated by DFO, would "wet the salt" the fish holds of just 16 Beverly schooners of the size that fished there during the 1850s.

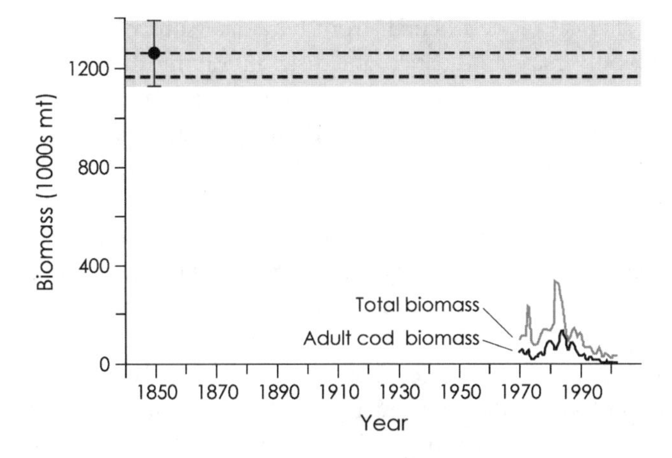

FIGURE 6.7. Biomass estimates for Scotian Shelf cod. This study (*solid circle in broken line*), with confidence interval, for 1852; estimated carrying capacity (*bold broken line*) of this marine ecosystem from late twentieth-century data; total biomass estimates (*gray spiked line*) from 1970 to 2000 for cod, 4X,4VsW; biomass estimates (*black spiked line*) from 1970 to 2000 for age 5+ cod, 4X,4VsW. Since prevalent hook size in this deepwater fishery made landing smaller juvenile cod unlikely, the biomass estimate for 1852 is compared with the current estimate for an adult cod population. (Rosenberg et al. 2005).

The Past and the Future

Reconstructing the narrative histories of people engaged with the ocean strengthens the notion that humans have long been a key component of marine ecosystems. Both the U.S. Commission on Ocean Policy and the Pew Oceans Commission called for ecosystem-based management of fisheries resources. To succeed, ecosystem-based management must relate biodiversity, and species abundance and distribution, to the productivity of marine ecosystems. At the same time, it must admit human responsibility in degrading ocean and coastal ecosystems.

The biomass of many fish harvested for centuries probably follows the trend in abundance we described for Scotian Shelf cod. The current abundance of many varieties of edible fish and mollusks may differ by orders of magnitude from what existed before mechanized harvesting. This has important ecological implications, particularly with respect to productivity. Thinking historically suggests new avenues of research into important ecological questions like the nature, magnitude, and resilience of productivity.

Current ecological models benefit from new data sources that document change in biodiversity, abundance, and distribution due to human agency.

By estimating the biomass of cod 150 years ago, we have challenged the current standard of a rebuilt cod stock in a productive marine environment. Contemporary disputes over the management of George's Bank and the Gulf of Maine often focus on standards for rebuilding fish stocks and the productive potential of marine ecosystems. Although some argue that a "fully rebuilt cod stock" need only reach 1980s abundance levels, both history and biology indicate otherwise. Management standards reflecting short-term political expediency fail to reflect the health of a cod population continuously fished for centuries. Rather, these standards reflect cod in decline and a profoundly distressed marine ecosystem. Our work on the Scotian Shelf cod fishery from 1852–1859 revealed a fishery, a marine environment, and a cod population significantly richer in size and structure than today's counterparts.

Contemporary regulators and fishermen have known for years that fewer fish exist now than "in the good old days." But the magnitude of the overall decline, which is the result of generations of shifting baselines, can only be revealed through systematic historical investigation and careful quantification of historic catches. It would not have been possible to estimate historic cod biomass without understanding the testimony of historic fishermen, and likewise impossible to explain changes in the nineteenth-century fishery without acknowledging declining cod stocks. So what have we learned? Some fishermen in recent years have declared that cod are coming back. Even if this is true, our biomass estimate of Scotian Shelf cod in 1852 shifts the baseline backward and changes the context for these observations. Although sobering, it provides a fundamentally different scale—a historical scale—against which to measure progress to date.

Acknowledgments

This study was part of the History of Marine Animal Populations (HMAP) Project, with funding support from the Alfred P. Sloan Foundation, part of the Census of Marine Life (CoML). Additional support was provided by the Mia Tegner Foundation, Maine Sea Grant, New Hampshire Sea Grant, the National Science Foundation (HSD-0433497), the Office of National Marine Sanctuaries, the Gordon and Betty Moore Foundation, the Richard Lounsbery Foundation, and the University of New Hampshire.

We thank Jesse Ausubel, Jeremy Jackson, Daniel Pauly, Don Clark, Stefan Claesson, Kate Magness, Mike Routhier, and Sherry Palmer. Special thanks go to Rob Gee for finding the logs in the first place. The staff at the National Archives and Records Administration in Waltham, Massachusetts, the Peabody Essex Institute in Salem, Massachusetts, and the Penobscot Bay Marine Museum in Searsport, Maine, assisted with historical research. Our colleagues Andrew A. Rosenberg, Andrew B. Cooper, and Matthew G. McKenzie shared equally in the research and analysis and deserve equal credit for these results. However, any errors in this paper are our own.

Chapter 7

History and Context: Reflections from Newfoundland

DANIEL VICKERS, WITH LOREN MCCLENACHAN

John Crosbie, the Canadian minister of fisheries, stunned Newfoundlanders on July 2, 1992, when he announced the first moratorium on the northern cod fishery, bringing to an end one of the oldest and richest fisheries on Earth. I was then teaching maritime history at the Memorial University of Newfoundland, and the news caught me entirely off guard. Though aware of the problems in the cod fishery for some years, I had not realized that things were as bad as they were—or at least I was not certain enough to do anything about it. For almost fifteen years I had been writing a book on the social history of colonial New England's cod fishery that dealt well enough with the hard lives of fishermen but ignored entirely the possibility that the livelihood they pursued might have been in the long run unsustainable. I had concerned myself only with the way in which the profits from the fishery had been shared and had ignored entirely the process through which the fishery itself was vanishing. And although it seemed true enough that the degradation of the cod stocks had not yet begun before 1850, the period I was studying, I could not run from the feeling that I had been dealing with a problem of the second order. Remarkably, two of my most talented colleagues at Memorial—Rosemary Ommer and Sean Cadigan—were completing parallel studies of the Gaspé and of Newfoundland in the nineteenth century that also focused on the social relations of production in the fishery rather than the question of its ecological sustainability. Although there is no reason to be ashamed of any of these books for what they did achieve, we were in some real way fiddling while Rome burned.

Once the severity of the ecological problem had been acknowledged, however, the collapse of the cod and other fisheries seemed almost certainly to be an issue with a history that was very deep. Indeed, as an historian, it strikes me as strange that anyone would dispute this. There may be terrific difficulties in the path of trying to establish how the oceans were once populated, and it is plain that the sort of detail ecologists usually demand of current research will never be possible for the distant past. But the same is true of human history. Large areas of understanding have been opened up during the last half century in territory that historians once believed was truly unknowable, and some questions to which students of the human past have produced clear answers bear a striking resemblance to the problems marine scientists address.

Here is an example from historical demography. It was once thought that the history of family structure—the biomass and age-class structure of humanity—could only be told with some precision back to the period of the first national censuses around 1800. Prior to that time, anyone interested in generalizing about such issues as age at marriage or family size had to rely on literary evidence and anecdote. Then, in 1963, Peter Laslett of Cambridge University published a short case study of family structure in a pair of seventeenth-century English villages, Clayworth and Cogenhoe, which undertook to measure some of these basic demographic phenomena. A colleague had drawn his attention to two documents dated 1676 and 1688, in which the parish rector of Clayworth in Nottinghamshire had listed for both years all the inhabitants of the village—around four hundred people—by household, occupation, relation to the family head, and number of times married. Although this document had been well known for many years, both to local scholars and to specialists in the history of Stuart England, nobody had ever considered before that a source of this sort might be used to reconstruct the population history of this particular village and that, indeed, if all the parish priests in England or "even some of them" had kept similar lists, "then the task of the historian of social structure would be transformed." By historians' standards, therefore, the stakes were pretty high. With even a generous handful of such lists, Laslett believed, "we should have the chance of reconstructing the population of our country as it was during all those generations which went by before the census began in 1801, of doing it swiftly, accurately, and completely."

Even with the little he had, Laslett discovered a number of astonishing facts. One was the extraordinary mortality and the constant reconstruction of households that was, in this seventeenth-century village, a fact of life; another was that households were a lot smaller (4.0–4.5 members) than Laslett had expected; third was the predominance of nuclear families; fourth was the ubiquity of servants, even in households of the middling sort; fifth was the unexpectedly large number of youths and children; and sixth was the high rate of mobility in and out of town. In short, the seventeenth-century, preindustrial English family looked a lot more modern than historians had thought.

Sharing with most historians the cautious spirit of a highly traditional discipline, Laslett did not push his generalizations very far. Amusingly for our purposes here, he likened his experience to that of a marine ecologist, "in his bathyscope [sic], miles beneath the surface of the sea, concentrating his gaze for a moment or two on the few strange creatures who happen to stray out of the total darkness into his beam of light." Were these communities typical of early modern England?

> On this the historian can only talk as the scientist might. Here are two samples of communities in motion, two tiny globes of light disposed at random a little way down into the great ocean of persons who lived and died in our country before records of persons in general began to be kept. These samples may be ordinary enough, but they may be quite extraordinary. We cannot yet tell: we may never be able to tell.

In reality, Laslett may have been a little disingenuous, for even as he wrote these words, he was at work on a much broader study, published two years later as *The World We Have Lost*. Meanwhile, a team of historical demographers, also at Cambridge, had discovered and begun to analyze several hundred similar lists dating from 1574 to 1821, and in 1981, the so-called Cambridge Population Group published the results of this project in the monumental *Population History of England, 1541–1871: A Reconstruction*. Laslett's work was hitched to a larger project—one of demonstrating that the entirety of the last five centuries in English social history was one of deep continuity and that the Marxist account of how the "horror of industrialization" had torn apart the nuclear family from the kin group with its "more humane, much more natural relationships" was simply a myth. Although this bolder claim is complicated and still a matter of debate, the bedrock of historical demography upon which it was based is now the stuff of textbooks. Early modern English families, as Laslett posited in

"Clayworth and Cogenhoe," suffered from high mortality; they tended to be small, young, and fairly nuclear in structure; and they moved around a lot.

There isn't an exact parallel between the Cambridge Population Group's research programs and those of the marine ecologists who gathered at the Scripps Institute of Oceanography in 2003 for a conference on Marine Biodiversity: The Known, Unknown, and Unknowable. Nobody in 1965 was afraid that the modern English family was on the verge of extinction, so there was nothing of the urgency inflecting the discussion of marine biodiversity at the beginning of the twenty-first century. In several senses, however, the two projects do bear comparison. Both deal with questions of considerable significance about transformations whose origins may well be buried in the distant past. Both face a paucity of evidence for the early periods and have long been prevented from addressing those questions in the temporal depth they demand. Both must build on a welter of models and assumptions to stretch these shards of evidence and data into a convincing portrait of the past. And neither will ever approach in completeness or certainty the standards of evidence that we demand of research conducted into families or oceans of the present day, where we as researchers have so much more control over what we choose to measure. The experience of historical demography, however, is an encouraging one. With little more basic data than would fit onto a couple of now antiquated floppy disks, scholars rediscovered a real part of "the world we have lost" in sufficient detail that we can now claim to have a powerful historical perspective on the contours of modern family life—one robust enough to allow us to avoid romanticizing a mythic past and imagining transformations that did not occur. Marine ecologists can reasonably hope to achieve the same.

The Panel Discussion

Why is the past important? In grappling with that seemingly basic question, I now return to a panel discussion on the subject at the Scripps conference. In many ways that conversation between Paul Dayton, Robert Paine, Michael Orbach, Alexander Stille, and myself with the audience characterizes a critical, ongoing debate about the relationship between history and ecology.

Paul Dayton argued that marine ecologists conduct their research too frequently on scales that are both geographically and chronologically small. The most appropriate regions in his view are large-scale basins or, possibly, an entire continental shelf, and the most useful timescales run in hundreds of years, when the distinction between cyclical and linear changes can really be tested. Dayton also warns against the pipe dream of trying to restore damaged ecosystems to some previous, supposedly desirable, pristine state. The Steller's sea lion population of the Bering Sea, for example, has collapsed under a variety of forces. Climate change, the reduction of food base by heavy fishing, and increased predation by killer whales are all hypotheses suggested to explain reductions in sea lion numbers. While past events such as commercial fishing and hunting of the great whales can be elucidated and even quantified, a direct link between past actions and the current state of the ecosystem has not yet been demonstrated. Each proposed hypothesis is overly simplistic and does not sufficiently explain recent sea lion population trends. The manner in which multiple factors have combined to alter the polar marine ecosystem is highly complex and as yet poorly understood, but the changes they have wrought are permanent, and no amount of human engineering is likely to bring the sea lions back.

Robert Paine, by contrast, was deeply skeptical about the prospect for understanding any ecosystem as large as the Bering Sea or any span of time as long as a century. Biological uncertainty coupled with multiple causation makes it hard enough to develop ecological generalities for places as small as Tatoosh Island in Washington State, over periods as short as the last few decades. Coupled to human impact are the background challenges of catastrophic die-offs, disturbances, and externally forced biological phase-shifts; how to sort these factors out over regions and periods as great as historical ecologists demand presents problems that Paine feels are insurmountable. If the past was as biologically variable as the present is, little can be learned from it.

Michael Orbach made a plea for incorporating the human cultural past into our understanding of modern ecosystems and especially the project of ocean conservation. The "tragedy of the commons," he argued, needs to be understood in the context of human expansion across the seas over the past millennium—charting and beginning to exploit the spaces and resources of most of the world's oceans, at least to the depth of a few hundred fathoms. With this expansion, there arose in the seventeenth century the doctrine of *mare liberum*, the freedom of the seas, under which most uses of the world's oceans remained unregulated within any common community except for the constraints of individual nation-states upon their own citizens. This happened for a very practical reason: no single nation or group of nations could effectively either monitor or control activities on the oceans, except within fairly close proximity to land. Accordingly, in the course of intense state competition over global commerce and marine resources, *mare liberum* emerged as a negotiated compromise. Since then, nation-states have gradually been advancing their claims over the resources of the continental shelves, arriving in recent decades at the establishment of 200-mile limits off certain coasts. Even more recently the world as a whole, through the United Nations, has at least begun to talk about a Common Heritage of Mankind approach to the problems arising from the overexploitation of oceanic resources. Yet, even today most of the world's salt water remains in a state of unregulated access with tragic consequences for many marine species. Orbach pointed out that not all common property has historically escaped serious regulation. Nations have managed to regulate use of the atmosphere to some better effect—in areas of air pollution, trading rights, telecommunications, and plane travel—which suggests that the enclosure of the seas is at least possible.

Alexander Stille raised the possibility that the destruction of marine biodiversity around the world may be closely linked to the decay of cultural diversity. Noting that the greatest concentration of linguistic diversity occurs in remote or isolated parts of the world that also register unusually high levels of species diversity, Stille drew our attention to the market-driven process of globalization and its homogenizing effects on the entire environment. Originally, this was driven by explorers and traders, then by colonizers and capitalists, yet now even well-intentioned conservationists have occasionally forced Western models of preservation such as the national park onto cultures where they make little sense. The rapid industrialization of the now-developed world has generated a connected but unrealistic cultural notion of pristine nature—uncontaminated by human contact—that has made many environmentalists hostile to the human community, complicating the process of addressing the real issue, which is the decline in diversity of all stripes—human or otherwise. An environmentalism with an appreciation for human culture has a far greater chance for promoting species diversity than one that sees man and nature as separate.

For all these differences in emphasis and opinion, two critical and quite distinct themes emerged: the uses of the *past*—by which I mean the accumulation of evidence before us—and the uses of *history*—or the story of humanity. While sorting these two out was not something that the participants attempted, each theme merits discussion.

The Past

The past is the property of every science. Ecologists, economists, astronomers, demographers, as well as historians draw repeatedly from the record of events, social and natural, to identify patterns and explain them. In certain sciences—geology is one and ecology another—the fundamental task has always been not the construction of predictive laws, but the explanation of how things have got to where they are, and in this case the past is all we have to go on. Although the quality and quantity of data collected by humans usually declines the farther into the past one delves, the investigation of any dynamic system demands some time perspective, and scientists in a wide variety of fields have proved highly ingenious at designing ways of using pieces of surviving evidence to learn something about what has transpired in the human and natural worlds. One can shower doubt on their methods, but their success in extracting patterns of any sort from this research means that they must be tracking something real. What, then, are the legitimate problems involved in mining the past for ecological data?

Two problems that seem to weigh heavily on the minds of ecologists are the related issues of predictability and design. Can evidence from the past be assembled that is solid enough to underpin working models allowing us to predict the future of the oceans, and can we use those models to design the sort of ecological results we wish? Ecologists strain toward predictability—partly because they are scientists and partly because they want to be useful—but many of them are skeptical of data from the past, the collection of which they cannot entirely control. Carl Safina pointed out that much legitimate science already depends on the imperfect record of the past—plate tectonics being an excellent example—and that marine ecology could well proceed along the same path. Yet Robert Paine probably spoke for many when he replied that marine ecosystems are terribly complex and prone to serious localized disturbances for which historical evidence will rarely be available.

On a microscale, of course, Paine is correct. Historians have to define their topics broadly enough to incorporate sufficient evidence to answer the questions they pose, and this rarely happens in the case of one microenvironment, say a fishing village, for example, over centuries of time. Accordingly, they broaden out to include a collection of villages along an entire stretch of coastline, and this inevitably does dull in some degree one's analytical edge. Although two of the most important fishing ports in colonial New England—Marblehead and Beverly—sat only fifteen minutes apart by sail, they employed two very different methods of recompensing the men and drew from two mutually exclusive labor pools. Lumping them together for the purpose of analysis obscures these differences, yet for reasons of limited evidence, this may be the only way to say anything significant—less than precise but generally true—about fishing society in

Massachusetts at all. Similarly in historical ecology, what one loses in precision by studying a large area such as a continental shelf, one may gain in significance through the power of generalization. The result may be imperfect, but if the big picture matters, it needs to be examined.

The New England cod fishery of the 1850s provides just such a case. Displaying no obvious interest in the separate ecosystems that probably made up the region confidently described as the single Scotian Shelf, University of New Hampshire ecological historians have agglomerated a great many imperfectly related data points into one set aimed at one question: how large were the cod stocks of that area in the age before the development of industrial fisheries? Their strategies will never allow them to penetrate the workings of that ecosystem in the way that some scientists might wish. Yet simply to know that cod stocks once stood at levels orders of magnitude greater than they stand today says something blunt but deeply significant about the changes that have transpired there over the last hundred and fifty years.

The root of the disagreement here would seem to hinge around the issue of predictability. Most will accept that we can discover something about the past—a series of rough benchmarks about marine populations before the advent of industrial fishing, for example. Whether or not we can use this information from the past to construct models reliable enough to enable us to undo the damage that humans have wrought in any precise way—"bending time's arrow backward," as Richard Hoffmann put it during the discussion—is a thornier issue. Ecosystems are too complex, and the models appropriate to describe them are too open-ended with too many unknown or unacknowledged externalities, to allow for anything like the sort of precision one would need to rebuild ecosystems as they were. Changes in the environment and consequent extinctions and near-extinctions in different seas have changed the pattern of species interaction forever and made certain waters uninhabitable to life-forms that once called them home. Were we to try to restore such complex environments under altered and imperfectly understood conditions, we would almost certainly produce unintended consequences. Though it is important to have some notion about what a desirable ecosystem might consist of, we must admit that any attempt to hit that target precisely will probably go awry and that a degree of caution born of humility must be invested in the reconstruction process, just as it ought to have been employed in the original march of destruction.

If there are advocates for predictive knowledge of the sort that would allow one to micromanage one particular coral reef or estuary back into some known, pristine state, they are not likely to be found among marine scientists and historians. The call is, rather, for a series of general quantitative measurements over wide areas that would alert us in an approximate sense to (a) what the carrying capacity of the oceans once was and (b) what the pace and direction of change in biodiversity and biomass, especially since the beginnings of measurable human impact, have been. If this seems simplistic, it is a point that many stakeholders in ocean resources still do not accept, and even to demonstrate roughly what (a) and (b) may have been would mark an advance over the present confusion.

In the general discussion that followed the panel, several participants connected the lack of awareness of the degree to which marine biodiversity is in decline to an increasingly ahistorical spirit abroad in the modern world. Carl Safina expressed a worry that in the twenty-first century the past was ceasing to have significance and that without any sense of it we would be "flying blind." Jeremy Jackson spoke of a "collective amnesia" that allows policymakers to commit the same errors over and over again without any clear awareness of how similar courses have proved disastrous before. Ram Myers offered a remarkable, graphic illustration of how the collapse of the northern cod in Atlantic waters at the end of the 1980s had been almost exactly foreshadowed by the crash of the haddock population in the same waters a decade before. Others echoed Daniel Pauly's concern that our baseline definition of a pristine ecosystem seems to have fallen with each successive generation.

While all of this is probably true, it is not clear to me that the problem is rooted in a general ahistoricity peculiar to the modern condition. Indeed, the very notion of modernity implies that we distinguish ourselves from a past that is qualitatively different from the present; indeed, we are bombarded daily with facts and myths from the past in a profusion that would have deluged our ancestors. Alexander Stille reminded us that even as archaeological monuments decay before our eyes, we feel prompted to record more information about them and try to learn more about how they came into being than earlier generations ever have. People in the past may have possessed more respect for tradition than we do, but that does not mean they had a clearer understanding of history—indeed, quite the opposite. They honored tradition because it spoke to them of things they felt had *not* changed. The problem is less that they were more conscious of history than we are and more that in a world that was changing much more slowly than ours, they could afford to be ignorant of history. We cannot. Roger Bradbury pointed out that it is not possible to see in the ecological history of past centuries anything close to what is happening now. The world is moving, as he put it, into a nonanalog state. Even though, within the human realm, the past has never been a very effective analog for the present, his broader point that the acceleration of change may be starting to outstrip our ability to understand and react informatively to environmental change is deeply worrying and demands that we try to chart out what these changes have been.

History

That the past, and sometimes the distant past, is relevant to the study of ecosystems should hardly require an extended defense. The touchier issue, and one that draws comment repeatedly, is the call for the integration of humans into the models of marine ecology. Integrating the *past* into ecology is one thing; integrating *history* into it is another. The latter might strike the uninitiated to be as self-evident as the former. We are certainly the top predators on our planet. Yet food webs are regularly plotted in diagrams where humans are not assigned any trophic level whatsoever; rather, we seem to enter into ecologists' current models of their ecosystems principally as an external force. Although we are still aliens below the waves, I doubt that after fifty years of dragging the continental shelves for fish, our otter trawls can still be considered in any meaningful way external to the struggle for life in the deep. Still, the consequences of incorporating *Homo sapiens* into marine ecology are complicated, especially in our role over time, and we need to think hard about what this may involve.

The central problem in letting humans into the story is that they possess consciousness and culture, both of which differ across space and change over time, making us very difficult to incorporate into any scientific model—ecological or otherwise. Because we as a species possess a degree of instability in our basic properties that manifests itself in the cultural time of decades and centuries, not in the evolutionary time of millennia and more, we are a shifting sandbar on which to construct theory.

Historians deal with this problem all the time. On the one hand, they consider it their job to reassemble the human past. On the other hand, they believe that absolutely all evidence from the past is in some sense corrupted three times over: through the processes of selection (what was recorded), preservation (what has survived), and analysis (what we choose to look at). Every document appears to them as if it were an object viewed through a series of wavy lenses, in which the object is the reality of something that went on in the past, and the lenses are the cultural filters that people of all the intervening generations have employed in their attempts to record, preserve, and interpret it. Part of the time, they are trying to use what they know about the different lenses to determine something about the object—the *behavior* of the past. And part of the time they are trying to use what they know about the object itself to study the lenses—the *culture* of the past. A historical document is, therefore, like a painting. It tells us something about the painter and something about the scene being painted, but for distinguishing between these two different somethings there is no scientific method. Becoming an historian, then, does not mean mastering a body of theory or a definable methodology. Rather, it involves learning enough about the practices and culture of a given age that one can look through a piece of surviving evidence *either* for the values of the day imbedded in the sort of story its author was trying to tell *or* for the real behavior one can discern behind the cultural presuppositions of the age. This may strike nonhistorians as a circular method—so it is—but it is inherent in a discipline that contends we are in some measure subjects of history ourselves. To this degree, history is an art and not a science.

In practice, therefore, historians will hardly ever be able to provide ecologists with the hard data they desire. Take an example from New England's woods and rivers. When the Puritan settlers first arrived in the seventeenth century they viewed most, if not all of what they saw—the salmon runs, the park-like forests, and the stocks of game—as features of an untouched wilderness. They saw it this way, however, only because they believed the Algonquian inhabitants of this land to be just as wild as the natural world around them and quite incapable of modifying it in any productive way. Knowing that the Puritans were unlikely to recognize a type of fishery based on brush and stone weirs, a style of forest culture organized around calculated burnings, or a form of culturally limited hunting aimed at a strictly regulated level of subsistence, we must now discount the Puritans' claim to have arrived at the edge of an unimproved world. These colonists had encountered in North America forms of land use that did not correspond to the world of hedges, sheepfolds, barns, and manure piles that English people equated with civilization—and they did not recognize the imprint that the Indians had already made upon the land.

Furthermore, they had a vested interest in seeing things this way. As the first governor of Massachusetts, John Winthrop, put it in 1629:

> That which lies common and hath never been replenished or subdued is free to any that will possess and improve it. . . . The natives in New England, they inclose no land neither have they any settled habitation

> nor any tame cattle to improve the land by, and so have no other but a natural right to those countries. So as if we leave them sufficient for their use we may lawfully take the rest, there being more than enough for them and us.

Before establishing their own right to the lands around Massachusetts Bay, Winthrop and his friends had to be convinced in their own minds that the landscape before them was unimproved. They also had to believe that the region was rich enough in natural resources to support them. Most of those early travelers to New England who took the time to describe the natural environment around them in forms (chiefly published) that have survived to the present were promoters, trying to persuade themselves and others that the Great Migration would work—that God really did intend this to be a refuge for his chosen people. For all of these reasons, writes William Cronon, one needs to take much of what they wrote with a grain of salt.

Nonetheless, their descriptions were more than simply self-justification and propaganda. All of these writers had visited New England (or spoke with those who had), and they were not just lying through their teeth. The indigenous peoples of New England had placed a fainter footprint on the land they had inhabited for thousands of years than the Europeans managed during their first two hundred. The early descriptions of alewives "pressing up in such shallow waters as will scarce permit them to swim" or shad "so thick . . . you could not put in your hand without touching some of them" plainly distinguish themselves from later descriptions of both these fish in vastly depleted numbers. There may not have been many foot-long oysters in banks that were a mile in length or clams "as big as a penny white loaf" when Europeans first arrived, but even if there were just a few of these monsters, or even if they were only nine inches long or half the size of a loaf of Wonder Bread, we know that by the end of the nineteenth century, mollusks were a lot smaller and much fewer in number than they had been at the time of European settlement. If one takes such anecdotal evidence in quantity, one can produce a picture of where and in roughly what quantity many marine species once were found that transcends the peculiar biases or mistakes of individual sources.

There remains, nevertheless, the difficulty that in all cultures, people are predisposed to see things or miss them, count them or ignore them, in consistent patterns we cannot disregard. Thus, the fact that in 1630 William Wood and many like him saw wilderness and abundance in the same places where by 1850 Henry David Thoreau and his generation saw industrial development and the beginnings of scarcity tells us something, not only about declining biodiversity in New England, but also about the growing sensitivity to that decline among New Englanders, and it may well be that comparing the two positions at face value may simultaneously exaggerate the degree of ecological transformation while it underestimates the amount of cultural change. This is no reason to flee from comparisons over time—in ecological history or cultural history—but an historian has to enjoin caution. That such evidence be used to demonstrate gross changes in different ecosystems over approximate periods of time and thus to contextualize the real pace of current change seems more than reasonable, but to employ this evidence to construct models capable of allowing us to engage in prediction at any level of real precision is unrealistic.

Letting humans into the picture not only introduces problems of evidence; it also tangles the process of analysis and understanding. The models that fisheries scientists now use to assess marine populations are far more complicated than they used to be. In single-species-based management, these models already involve an attempt to monitor fishing mortality through measurements of fishing effort, effectiveness of gear, and size of catch at different age levels. Daniel Pauly has suggested that in ecosystem-based management, they might also include "trophic interaction between species, habitat impacts of different gears, and a theory for dealing with the optimum placement and size of marine reserves." In such models, humans appear chiefly through their effect on gear and effort, which is probably also how the fish see it. But gear and effort are functions of much more complicated human systems that are in constant motion over time. Ecologists cannot study everything at once, of course, so they rely on economists, demographers, lawyers, and a few other specialists in human behavior to deal with these factors. This may work in the short run, if the social science is good enough, but in the longer run the possibility of cultural change enters the picture and foundations upon which the social scientific models are built begin to buckle.

A classic and fairly simple example is the relationship between prices and effort. At the dawn of the twenty-first century, it seems clear that in all commercial fisheries these are positively related and that, other things being equal, rising fish prices will prompt increasing fishing effort and improvements in gear. While I know of no studies of preindustrial fisheries that address this relationship, one work on the fur trade in western Canada during the eighteenth century and another on family-run mining operations in early modern England demonstrate the opposite—that in both cases, rising prices brought about diminished effort, as producers capitalized on improved returns by maximizing not profits but leisure or at least the time to invest in alternate economic activities. An economist has no trouble describing these responses; they both take the form of a classic backward-sloping labor supply curve. But the shape of the curve is culturally determined—a historical creation—and its history never has been or ever will be subject to prediction.

A more complicated cultural question—and one that has attracted much interest among maritime historians, anthropologists, sociologists, and others—is the concept of the common property resource. It is undoubtedly true that most of the ocean has been treated this way, especially in the last five hundred years as Michael Orbach pointed out. Building on this truth, economists and biologists have generated predictive theories of some power and considerable influence. Yet it is also true that there have been across the ages few utterly unregulated commons. All resources are exploited along a spectrum of legal and customary environments, each of which contain a mix of public and private privileges that have developed historically. Some commons have degraded very rapidly, others have survived much longer, but all of them have a history, which is basically one of changing human assumptions about what one can and cannot do with the resource. The indigenous salmon fishery of British Columbia, for example, was sustained for something in the order of three thousand to five thousand years before European settlement on the Pacific coast, not because Indians lacked the technical capacity to overfish or because they fished for personal consumption only, but because they possessed an economic culture that emphasized subsistence, diversification, and reciprocity and because their tribal culture was not inherently expansionist. When the Europeans arrived in the nineteenth century, they introduced a different economic culture—based around commodity production, specialization, and seizure of the land for settlement—that succeeded in destroying the salmon stocks in many places along the coast within less than a century. This particular tragedy of the commons is not an expression of an economic law but a complex historical tale. No model constructed to explain the ecosystem of the Pacific coast in 1800 (even with the best science) could have foreseen what has transpired since. Allowing humans and the historical process into the picture adds a measure of instability into our understanding of the environment that does not obviate the need for science but certainly clouds the prospects for prediction.

The trouble with introducing humans into ecology is not that it is unnecessary—quite the opposite. The problem is rather that it ushers in a series of variables changing in nonlinear fashion that we cannot study with true scientific objectivity because they are extraordinarily complex and because they are us. The elements of culture that can shape a given ecosystem may spring from areas and activities that seem at first glance to have nothing whatsoever to do with the environment in question. Always they are rooted in events that transpired decades or even centuries before the degradation of that ecosystem began. And they often work at cross-purposes, pushing people to take some actions that protect the environment while simultaneously doing other things to ruin it. In the short run, the decisions that humans make may seem like pure stupidity—"the march of folly," to borrow Barbara Tuchman's phrase. But they are also embedded in cultural assumptions of enormous power that we cannot easily, even humanely, throw aside.

Newfoundland Redux

Which brings me back to Newfoundland. Here we have what Carl Safina called with some justice "the worst fisheries management failure in the world." What happened there in the second half of the twentieth century might well be described as a species of criminal imbecility, but it was also rooted in cultural forces of great power and much antiquity. Those who were connected to the tragedy—if only as passive observers like myself—must accept responsibility for what happened, but we cannot trivialize the difficulty in seeing things for what they are when we ourselves are implicated in so many different ways in the culture that does the damage. Although I am not an historian of Newfoundland myself, permit me some general reflections on what those cultural forces may have been.

The first of these, oddly enough, is a form of nationalism. Newfoundland was settled for the most part between 1775 and 1815—an extraordinarily short period of time when recurring maritime warfare destroyed the migratory fishery from Europe and high fish prices enabled thousands of people to move permanently to the island. Those conditions disappeared with the Treaty of Ghent in 1815, leaving a population of about forty thousand to make their way in a place where almost everyone lived on the knife's edge of survival. For the better part of two centuries in the face of constant hardship they managed, scraping by on erratic returns from a fishery financed by merchant capital. And in the course of time Newfoundlanders developed their own dominion government, a powerful sense of their own identity, and a nationalist feeling that this was a place where they should be allowed to live.

The decision to join Canada in 1949 as its tenth province constituted something of a devil's bargain, whereby Newfoundlanders gave up considerable authority to Ottawa in return for the latter's promise to help finance their continued occupation of the island. This financial assistance has taken innumerable forms since confederation: support for capital improvements inside the fishery, development of infrastructure across the island, funding for medicare and higher education, and a generous program of unemployment insurance that functions as a form of income supplement for fishing families. Newfoundlanders have always been willing to emigrate if necessary, but their commitment to the island is powerful enough that they would rather not, and after 1949 enough of them bought into Ottawa's commitment to develop the province's fishery that they eventually helped destroy it. Sumaila and Pauly write with some justice of the disastrous ecological consequences of subsidizing the fishing industry and the communities that pursued it. Yet it is hard to imagine how in a democracy there would have emerged the political will to renege on the bargain of 1949, squeeze Newfoundlanders out of their modest homes, and crush their cultural traditions. This was not simply stupidity but also a considered response to real moral issues and a nearly intractable historical problem.

The second cultural force of consequence was Newfoundland's political weakness. As a colony and dominion within the British Empire, the island's interest in the marine ecosystem that surrounds it was always subordinated to Britain's wider imperial concerns. On several occasions in the nineteenth century, prompted by complaints over poor catches, the Newfoundland legislature did attempt to take measures to regulate its fishery, only to be told by London that these were matters of maritime and, therefore, foreign policy, over which the island had no jurisdiction. When Newfoundland joined Canada, it hitched itself as a minor partner to a profoundly continental country—born of the fur trade into the interior and knit together by the building of the Canadian Pacific Railroad—*a mari usque ad mare*, but no further. To this day, the effectively populated portion of Canada retains a ribbon-like quality, 3,000 miles long and a couple of hundred miles wide. Most Canadians live far from the ocean and do not think of it very seriously if at all. In times of crisis, Ottawa has intervened (as, for example, to establish the 200-mile limit in 1977 and to close the cod fishery down in 1992), but most of the time the fishery has been characterized mainly as a problem—a "failed staple," an "employer of last resort," or a "chain with too many weak links"—and the project of dealing with it constructively has never managed to attract the imagination of Canada's ruling class or the majority of its voters.

The Canadian case is not unique. The advent of the industrial revolution and the arrival of the railroad throughout the world in the nineteenth and twentieth centuries focused human attention toward manufacturing processes and the profits to be won by colonizing the interior of the great continents and developing their landward resources. Maritime events, which had been at the center of world history since the beginning of the fifteenth century, ceased to be so after 1815. So while the exploitation of marine resources has intensified in the last two hundred years, the management of the seas has become largely the responsibility of nations possessing a maritime consciousness that is stilted at best. This is not to say that in preindustrial times, the Dutch, French, or English states were ecologically conscious. But at least their fisheries figured prominently in national policy, which is more than one can say of most countries today.

Finally, and most profoundly, the destruction of the northern cod cannot be understood apart from the historical construction over the past five centuries of a global marketplace. Throughout this period, the freedom of whalers, sealers, and fishers to descend on virtually any marine population with the full force of human technology, unless specifically prohibited by particular laws, has been the general rule. During the last fifty years, many governments have grown more aggressive about reserving coastal waters to their own fishermen, yet the default position in capitalist culture is that whatever is not explicitly protected is fair game. In the months that followed the cod moratorium in 1992, the Newfoundland media and general public thrashed about looking for an explanation of what had happened. Some named the Department of Fisheries and Oceans as the culprit and others the politicians, some the foreign fishermen and others the seals. A good many blamed themselves: "We took too much," some of my neighbors told me. But almost nobody blamed National Sea Products or Fisheries Products International—the two corporations who in the name of the bottom line engineered most of the final slaughter. Their behavior was held to be completely natural and no more culpable than the weather. To this day, neither company displays much visible remorse over its role in driving the northern cod to the edge of extinction. FPI's website reported only that "the Newfoundland groundfish industry has faced a number of resource challenges in recent years," a piece of bland corporat-ese that reflects the company's belief that since it did nothing illegal it was not responsible for the disaster. It was only servicing demand.

If the right to catch cod, and most other fish, is lightly regulated today, the right to consume fish is freer still. Nations may use tariffs to protect local fishers from foreign competition, but rarely will they regulate consumption per se, especially if they possess no fishery of their own. Until internationally recognized as an endangered species, there exist almost no rules to restrict the trade and consumption of legally caught fish. For the most part, we simply expect market forces to drive the price of this scarce commodity upward and push people into searching out other species or entirely different sources of protein altogether.

Limitations on consumption are not unknown in history; dozens of different cultures have followed dietary restrictions, for example, that forbid certain foods outright. In the modern West, however, we have enjoyed many centuries of prosperity derived largely from market exchange and are deeply vested in the freedom to buy and sell the commodities we produce as well as the factors of production we need to produce them. Fundamentally, it is what distinguishes us from serfs and slaves. Now and again, especially during periods of crisis (wars spring to mind), we will accept limitations on these freedoms, but our tolerance for them is normally very low. Especially in the United States, but to some degree in most Western countries, it is extraordinarily easy to protest any specific economic restriction on consumption—no matter how important it may seem—by appealing in general terms to the principle of liberty. So fully are we implicated in the culture of choice—learned over centuries of coping with capitalism—that it is difficult to mount any broad campaign to persuade one another to adopt restraint. We will *have* to learn, but undoing five hundred years of being taught the opposite will not be easy.

That we need to study the past both to understand what is happening to marine ecosystems and to manage human involvement in them today should go without saying. We cannot, however, underestimate the difficulties involved in this project. Integrating ecology and history will force ecologists to confront culture, a phenomenon that complicates the collection and interpretation of evidence and that evolves in unpredictable ways. Insofar as cultural change occurs gradually and in knowable directions, we can allow for it in our attempt to reconstruct ecological behavior in the past—but rarely with the sort of precision a scientist would like. Understanding the rootedness of human ecological behavior in history helps us to see how we got to where we are—but it also alerts us to the many ways in which the values we inherit set the switches for the decisions, some of them quite disastrous, that we make. We are not completely the prisoners of those values;

if we were, nothing would ever change. But we call them values because we value them, and we err if we think that casting them aside is a simple matter.

Acknowledgments

Panelists Paul Dayton, Michael Orbach, Robert Paine, and Alexander Stille contributed to the 2003 discussion from which this chapter evolved. Their ideas and those of others who engaged in the conversation are included here, with thanks.

PART IV

Methods in Historical Marine Ecology

Here the focus shifts from specific research questions and results to methodology. Heike Lotze and colleagues review the remarkably broad array of tools and analytical approaches that have been used with varying degrees of success to reconstruct the ecological history of the ocean. Then Stephen Palumbi addresses the question of how many whales existed in the past before we began to kill them and why different kinds of analyses can yield strikingly conflicting results. The "devil is in the details" in historical marine ecology as in any other discipline.

Analytical methods are so varied that large teams of different kinds of specialists are often required to assemble a thorough picture of past events. For this reason historical ecology is fundamentally collaborative and synthetic. Lotze and co-authors present examples of techniques gleaned from paleontology, geology, radiometric dating, stable isotope chemistry, archaeology, genetics, history, early and modern scientific surveys, and ecological experiments, among others. All these fields are potentially relevant, and several recent review articles employ data from most of them. The trick, of course, is to know the strengths, weaknesses, and assumptions of different methods, to judge which ones are more likely to bear fruit under relevant conditions and over different temporal and spatial scales.

Sometimes it is enough to estimate the abundance of organisms or environmental conditions at some time in the past as a frame of reference to compare with the present. But in other cases, we need well-constrained time series of data to document long-term trends in ecosystems or establish relationships of cause and effect. With better baselines and longer trajectories of change, managers can better anticipate the likely environmental consequences of different actions.

In the happy case when the results of different methodologies agree, confidence increases in our picture of the past. A good example is the remarkable correspondence between historical estimates of past abundances of cod on the Scotian Shelf versus calculations based on the carrying capacity of the environment and the biological characteristics of cod. But in other cases, different methods can produce drastically different results, as Palumbi describes for estimates of pristine abundances of whales. Calculations based on historical catch data differ substantially from those based on genetic analysis of mitochondrial DNA sequences. The historical estimates appear too low because most of the assumptions used in the genetic analyses err on the conservative side whereas the historical records are inevitably incomplete. The situation is still unresolved.

The controversy about whales is of more than academic interest because estimates of pristine whale populations are used by the International Whaling Commission (IWC) to determine the health of whale populations today. Low population estimates before whaling would indicate that some species, such as minke whales, are close to recovery. High population estimates before whaling, on the other hand, would mean that whale populations today are still severely depleted and cannot support commercial whaling. Because Japan and Norway want the IWC to lift the moratorium on commercial whaling, which method is more accurate is bitterly debated and has direct bearing on this controversy.

The same is true for virtually all overharvested fisheries and for any other environmental problem for which modern baseline data are lacking. Much work needs to be done to establish guidelines for the application and interpretation of different kinds of historical records and to determine which ones to use under the conditions at hand. Clearly stated assumptions and transparent data are vital. The more different methods that can be brought to bear on reconstructing the past, the more convincing will be the results when they agree and the stronger the argument for regulatory action.

Chapter 8

Uncovering the Ocean's Past

HEIKE K. LOTZE, JON M. ERLANDSON,
MARAH J. HARDT, RICHARD D. NORRIS,
KAUSTUV ROY, TIM D. SMITH,
AND CHRISTINE R. WHITCRAFT

The ocean has a *history*. Over the past hundreds, thousands, and millions of years the ocean and life within it have evolved and changed. This history determines the ocean's present state and shapes its path into the future. We cannot understand how the ocean functions today without knowing its ecological history. Likewise, we cannot predict future changes without knowing the origin, cause, and trajectory of change in the past. Finally, we cannot effectively restore degraded marine populations, communities, or ecosystems without historical baselines to use as reference points.

Marine historical ecology is a new field concerned with compiling and understanding the ocean's ecological history. It is a multidisciplinary effort to piece together a puzzle from fragments of the past. That *past*, however, was ever changeable so that our baseline for comparison depends on when we choose to measure it. Over time the baseline may have shifted due to natural variability, human impacts, or a combination of the two. Major goals of historical ecology are to determine the direction and magnitude of such changes, and to distinguish between natural and human cause and effect. In particular, we want to answer these questions: What has changed in the ocean? What caused these changes? And where do these changes lead us?

Answering these questions requires information from many different disciplines, including paleontology, archaeology, history, and ecology. The kinds of data available and their strengths and weaknesses vary, but they all provide valuable information about the past, which can be pieced together to reconstruct an ecological history. The process is fundamentally retrospective (figure 8.1). Paleontologists analyze sedimentary records for fossils and proxies of environmental change including isotopic composition, biochemical traces of organisms that are not fossilized, and in some cases ancient DNA. Archaeologists do much the same for artifacts, plant and animal remains, and patterns of human habitation and exploitation. Historians analyze early narrative descriptions and archival records of what people saw, caught, traded, and sold. Ecologists compare biological and environmental observations, surveys, and experiments in the past with contemporary data. Mathematical tools such as modeling, time series analysis, and meta-analysis help to compare and integrate all these different kinds of data into a more coherent whole.

Here we provide an overview of different research disciplines, data, and methods used to reconstruct the ocean's past. Then we highlight opportunities to integrate results in order to develop a comprehensive understanding of the past that stretches from deep time to the present. Our purposes are to envision the kinds of biological and environmental changes that have taken place in the ocean over different temporal and spatial scales, not just for individual organisms but for entire food webs and ecosystems, and to differentiate natural and human causes of change.

Disciplines and Data

Paleontological, archaeological, historical, and ecological data vary widely in their temporal and spatial scale, precision, and application to the historical ecology of the ocean (table 8.1).

Paleontological and Geological Data

The fossil record is the primary source of information about changes in species and environments over centuries and millennia to millions of years. Macrofossils (figure 8.1B) provide information about species that are extinct today, or about occurrences of extant species in the distant past. The sedimentary record represents archives of past communities and ecosystems and thus a record of ecological and environmental change. This is because layers of sediment are deposited over distinct periods of time, and the age

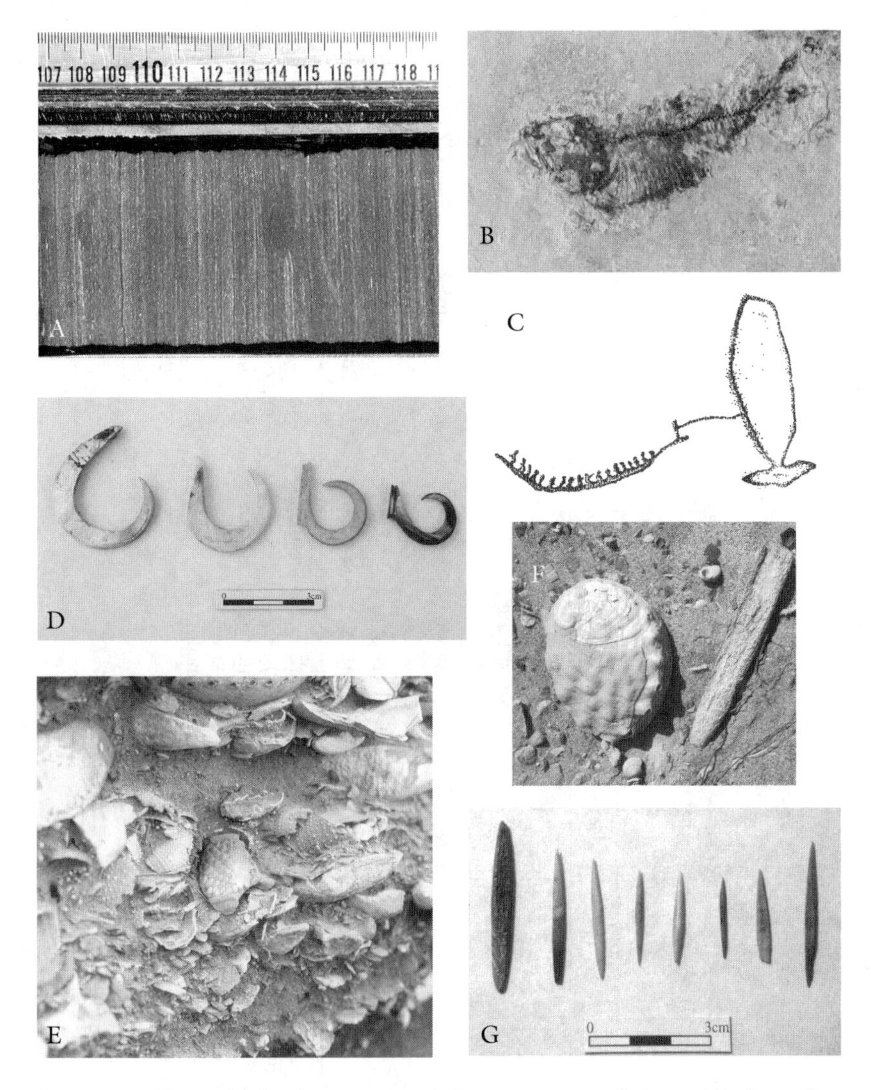

FIGURE 8.1. Examples for data sources used to reconstruct the past. (A) Core photograph of laminated sediments from Ocean Drilling Program Hole 1261A, core 41, section 1, Demerara Rise, NW South America (photo provided by Richard D. Norris). (B) Fossil sardine from laminated organic claystones in the Monterey Formation, Gaviota Beach, Santa Barbara, California, approximately 11 million years old (photo provided by Richard D. Norris). (C) Neolithic wall drawing on a sandstone wall at Ulchu in southwest Korea, showing a whale harpooned by a boatload of men (from Ellis [1991], image used with permission). (D) Circular fishhooks made from shells, less than 2,500 years old, San Miguel Island, California (photo provided by Torben C. Rick). (E) Shell midden site at San Miguel Island (photo provided by Jon M. Erlandson), (F) yielding animals such as abalone (photo provided by Jon M. Erlandson), and (G) archaeological remains of bone gorges, ca. 9,000 years old (photo provided by Jon M. Erlandson).

TABLE 8.1. Comparison of qualities, temporal scales and spatial scales, for data derived from different disciplines. For more detail, refer to text.

Data source	*Data type*	*Bias*	*Timescale*	*Resolution*	*Spatial scale*	*Example*
Paleontological data	O, R	B, E, M	short-long	high-low	small	figs. 2, 3
Archaeological data	O, R	B, E, H, M	short-long	med-high	med	figs. 4, 5
Genetic data	O, R	B, M	long	low	large	
Historical data	O, R, A	H, M	short-long	high	med-large	figs. 6, 7, 9
Monitoring data	O, R, A	H, M	short	high	med	fig. 8
Fisheries & hunting data	O, R, A	M	short-med	high	med	figs. 7, 9
Experimental data	O, R, A	H, M	short	high	small	

Data type: O = Occurrence (presence, absence), R = Relative abundance, A = Absolute abundance
Bias: B = Biological (depending on species, e.g., selective preservation), E = Environmental (depending on environment, e.g., selective preservation), H = Human (depending on activities, e.g., selective use / interest), M = Methodology (depending on analytical method)
Timescale: long (> 1,000 years), medium (100–1,000 years), short (< 100 years)
Resolution: high (yearly or less), medium (decadal to century), low (millennial)
Spatial scale: large (> 1,000 km), medium (1–1,000 km), small (< 1 km)

of each layer can be determined by a great variety of methods, including radiometric dating, the first and last occurrences of fossils, and climate records based on stable isotopes or pollen. Although most sediments are mixed to some extent by burrowing organisms, annually laminated deep-sea sediments (figure 8.1A), especially those in anoxic basins, can even preserve bimonthly environmental records spanning millennia. The relative abundance of isotopes and minerals in different sedimentary layers is used to reconstruct climate and environmental variability in the past. For example, the oxygen isotope ^{18}O is a common proxy to determine past temperatures and reconstruct paleoclimates (figure 8.2). Likewise, strontium and titanium can be used as proxies for paleosalinity and the amount of precipitation and river runoff over thousands of years.

In addition to the sedimentary record, isotopic and trace element analyses of skeletal growth bands of long-lived organisms such as some mollusks and corals can provide a record of past environmental conditions. Because corals grow incrementally every year, they show growth layers just like tree rings and may preserve seasonal climate cycles over intervals of several hundred years. The fraction of stable isotopes of O and elemental ratios such as Sr/Ca (strontium/calcium) record the environmental conditions when each layer was formed. For example, O measurements of long-lived shallow-water corals have been used to determine a 240-year record of precipitation and regional runoff in the Florida Keys. High-precision

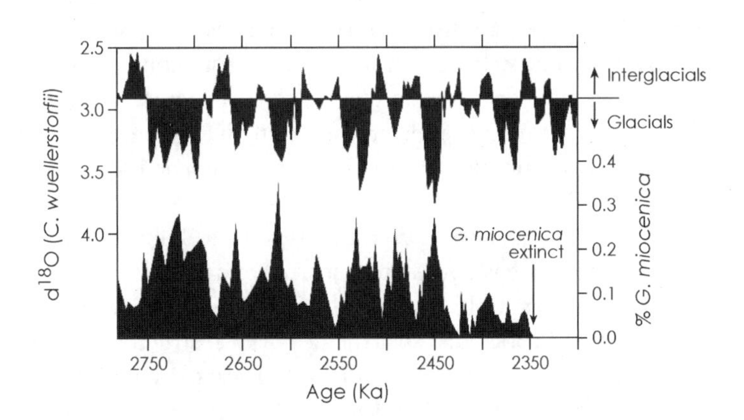

FIGURE 8.2. The sediment record: climate and plankton. Reconstruction of the 41,000-year cycle in abundance of the planktonic foraminifer *Globorotalia miocenica* in response to cyclical changes in global climate derived from oxygen isotope records from benthic foraminifera. (Adapted from Ocean Drilling Program Site 999, Caribbean Sea. Adapted from Norris [1999]; Haug and Tiedemann [1998]).

radiometric dating and the construction of master chronologies using characteristic patterns of growth can be used to reconstruct the environmental variability of ecosystems with daily to monthly to annual precision extending back thousands of years.

The fossil record is also the primary source of information about when and how rapidly species first appeared or went extinct, and how relative abundances of species and higher taxa changed over time. The fossil record shows that populations of many species display large fluctuations in abundance on a variety of timescales, which may translate into variations in susceptibility to extinction. For instance, the Caribbean planktonic foraminifer *Globorotalia miocenica* displays large variations in abundance between 2.8 million and 2.3 million years ago with the growth and decay of glacial ice, being more abundant during glacial than interglacial periods (figure 8.2). The abundance of this species repeatedly dipped low during interglacial periods and ultimately became extinct globally during one of these episodes.

Similarly, pelagic fish that shed their scales throughout life leave clear records of their past abundance in ocean sediments, as observed for populations of the northern anchovy (*Engraulis mordax*) off the California coast during AD 278–1956 (see MacCall, figure 4.2, this volume). Relative abundance fluctuated in cycles of about 102 years, a periodicity far longer than any biological observations available for the entire Pacific Ocean. This periodicity was not evident in existing catch statistics and underlines the importance of the sediment record in evaluating population dynamics of this commercially important fish. This begs the question of whether population fluctuations reflect changes in abundance throughout a species' range or shifts in geographic distribution that reflect environmental fluctuations (table 8.1). Distinguishing between these alternatives requires spatially explicit analyses using multiple samples along environmental and geographic gradients.

The fossil record can also be used to assess the stability of ancient ecosystems and changes in basic ecosystem structure that predate modern ecological observations. Pandolfi studied a succession of uplifted reef terraces on New Guinea and concluded that the population structure of reefs had remarkable stability over a period of ~ 100,000 years. A comparable study of uplifted reef terraces on Barbados showed that the widespread loss of the coral *Acropora* from the Caribbean in the 1990s had no parallel in the fossil record. Wapnick and colleagues arrived at a similar conclusion based upon the temporal distribution of coral rubble in sediment cores from Discovery Bay, Jamaica. In a different setting, analysis of the abundance and radiometric age of bivalves from the Colorado Delta was used to document the wholesale switch in species dominance associated with the reduction of Colorado River flows in the early twentieth century.

The sedimentary record also provides an archive of past human impacts such as increased sedimentation rate, pollution, eutrophication, and anoxia. For example, the Ba/Ca (barium/calcium) ratio in cores of large coral heads provides a proxy for sediment fluxes from river runoff, which usually increased after permanent settlement and land clearing. Cooper and Brush used a variety of proxies from sediment cores to demonstrate that water quality in Chesapeake Bay deteriorated quickly after permanent European settlement in the seventeenth and eighteenth centuries (figure 8.3). They calculated sedimentation rate by measuring concentrations of terrestrial pollen in each sediment layer, an index of eutrophication based on the ratio of centric diatoms and pennate diatoms, as well as the total content of organic carbon and nutrients. They also used the content of sulfur and occurrence of the iron sulfide pyrite as proxies for hypoxia and anoxia. Even simple comparisons of the compositions of living benthic communities with time-averaged death assemblages are a very good indicator of past anthropogenic change.

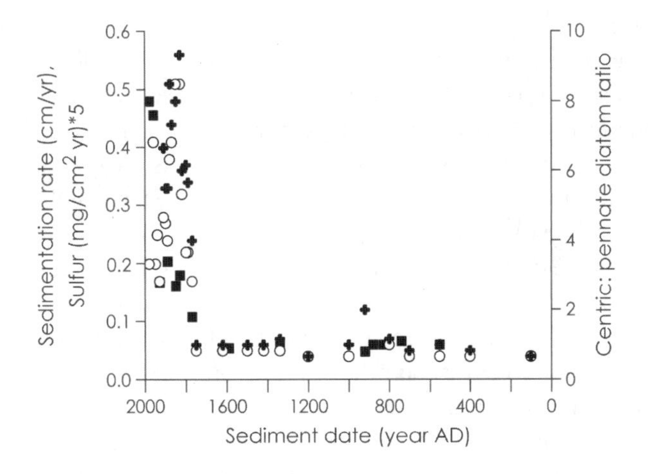

FIGURE 8.3. The sediment record: pollution. Reconstruction of anthropogenic effects on water quality of Chesapeake Bay over the past 2,000 years. Data from sediment cores indicate strong increases in sedimentation rate (*solid squares*), the ratio of centric : pennate diatoms as a proxy for eutrophication (*open circles*), and the content of sulfur as a proxy for anoxia (*crosses*) after European settlement in the seventeenth to eighteenth century. (Data adapted for Core R4-50, with permission, from Cooper and Brush [1993].)

The fossil record provides a picture of environmental and ecological change over very long timescales that can be invaluable for placing current changes into context so that we may better distinguish the natural variability of ecosystems from anthropogenic effects. However, the spatial and temporal scale of this record varies among regions and environments. Deepwater microfossils can provide high-resolution records of past environmental change on a global scale but tend to include only a small number of species. Shallower-water macrofossil assemblages are much coarser in temporal resolution, but are typically more diverse and include many commercially important groups. However, the record consists primarily of species that have preservable hard parts such as shells and bones, and some groups have higher preservation probabilities than others. Quality of preservation also depends on the environment of deposition and the extent of subsequent alteration. There are also problems in estimating relative or absolute abundance compared to simple presence or absence (table 8.1), although this is sometimes possible under special circumstances.

Archaeological Data

Archaeologists analyze remains from past human settlements that mirror former human activities, their tools and ornaments, as well as the species that were used for food, medicine, or other needs. Remains can be found in caves and graves or buried in the sediment of former occupation sites. Food remains in the form of animal bones and shells, as well as remnants of plants, are found in coastal shell middens (figure 8.1E) where they were deposited over time by subsequent settlement cultures. Different strata within such sites can be dated through a variety of techniques, including radiocarbon dating. Thus, archaeologists can use stratified shell middens or sediments as archives of past information revealing human technology (figures 8.1D, 8.1G), resource use and other cultural practices, and environmental parameters. Archaeological assemblages are culturally selected and biased by human activities, interests, and preferences. Nonetheless, middens commonly contain diverse biological assemblages collected by humans for a variety of purposes, including wild and domesticated species as well as stomach contents and epifauna associated with various prey (table 8.2).

The most common plant and animal remains found in middens or sediments are pollen, seeds, shells, bones, scales, teeth, and hairs. In addition, most archaeological sites also contain remains of tools and weapons that give clues about the technologies humans used for exploitation. Plant remains provide information on wild or cultivated plants used and the natural vegetation surrounding the settlement area. Animal remains reveal past occurrences of particular species, their use by humans as food or ornaments, and information on their size, age, and relative abundance. Over the past 11,000 years inhabitants of Daisy Cave on the Channel Islands of southern California consumed more than 150 species of marine and terrestrial animals, some of which are extinct today (table 8.2). The relative abundance of different species in various strata indicates changes in their importance as nutritional resources over time. Intertidal shellfish and near-shore fish were generally of greatest importance throughout the sequence, followed by mammals, while birds were of minor importance (figure 8.4).

From these and other records it has become clear that anatomically modern humans have heavily influenced coastal and island environments for much longer than previously believed. Marine shell fishing, fishing, and hunting are now known to date back to the last interglacial stage, more than 100,000 years ago. The colonization of Australia and western Melanesia 64,000 to 35,000 years ago documents the Pleistocene use of seaworthy boats, and there is evidence for the colonization of the California coast by about 13,000 to 12,000 years ago.

Some sites clearly reveal the effects of human exploitation on the relative abundance or size of species over time. Highly valued species such as sturgeon and geese in the Emeryville shell mound in San Francisco Bay declined in relative abundance (figure 8.5) and were replaced by smaller and less valued species, indicating resource depression by indigenous people. The average size and age of sturgeon as indicated by their dentary width also declined over time. In contrast, Atlantic cod (*Gadus morhua*) consumed along the Gulf of Maine remained nearly one meter long in size for 4,000 years until intensive commercial fishing by European settlers, suggesting that indigenous people may have had little impact on this commonly used species. However, a detailed analysis of archaeological records in Penobscot Bay, Gulf of Maine, revealed a distinct trend of decreasing apex predators (mainly cod and swordfish) and increasing mesopredators (flounder, sculpin) between 4,350 and 400 years ago.

Archaeological remains also provide information about species that occurred in earlier times but are now regionally or globally extinct (table 8.2). Excavations at several geographically distant sites help to reconstruct how distributions of species changed over time due to natural or human impacts. For example, breeding colonies of fur seals had persisted on both the North and the South Islands of New Zealand prior to the arrival of the Maori people in AD 1250. The Maori hunted fur seals for food. By 1500,

TABLE 8.2. Selected animal species identified in archaeological and paleontological strata from Daisy Cave, San Miguel Island, California.

Common name	*Identified taxa*	*Notes/Comments*
Marine mammals		
Gray whale	*Eschrichtius gibbosus*	
Common dolphin	*Delphinus delphis*	
Seals and sea lions	*Arctocephalus townsendi*, *Callorhinus ursinus*, *Phoca vitulina*, *Zalophus californicus*	Food, fur, oil
Sea otter	*Enhydra lutris*	Food, fur
Marine birds		
Bald eagle	*Halieetus leucocephalus*	Locally extinct
California condor	*Gmnogyps californianus*	Locally extinct
Albatross	*Diomedea albatrus*	
Pelican	*Pelicanus occidentalis*	
Cormorants	*Phalocrocorax pelagicus*, *P. penicillatus*	Breeding colony at site
Gulls, terns	*Larus californicus*, *L. occidentalis*, *Sterna paradisaea*	Migratory
Shearwaters	*Puffinus griseus*, *P. puffinus*	Migratory
Storm petrels	*Loomeliana melania*, *Oceanodroma domochroa*, *O. leucorhoa*	
Murres and murrelets	*Uria aalge*, *Endomychura hypoleuca*, *Synthliboramphus antiquum*	
Geese, black brant	*Chen hyperborean*, *C. rossii*, *Branta nigricans*	
Crane	*Grus canadensis*	
Ducks, teals	*Anas* sp., *Oxyura jamaicensis*	
Scoters	*Chendytes lawi*, *Melanitta deglandi*, *M. perspicillata*	
	Chendytes	Extinct
Grebes	*Aechmophorus occidentalis*, *Podiceps auritus*, *P. caspicus*	
Loons	*Gavia arctica*, *G. immer*, *G. stellata*	
Black-bellied plover	*Squatarola squatarola*	Migratory
Sanderling	*Crocethia alba*	Migratory
Fish		
California sheephead	*Pimelotopon pulchrum*	Major food species
Rockfish	*Sebastes carnatus*, *S. flavidus*	Major food species

TABLE 8.2. Continued

Common name	*Identified taxa*	*Notes/Comments*
Topsmelt	*Atherinops affenis*	
Monkeyface eel	*Cebidichthys vioaceus*	
Cabezon	*Scorpaenichthys marmoratus*	
Surfperch	*Cymatogaster gracilis*, *Hypsurus caryi*, *Embiotoca jacksoni*	
Pile perch	*Damalicthys vacca*	
Giant kelpfish	*Heterostichus rostratus*	
Lingcod	*Ophiodon elongatus*	
Sharks	*Mustelus californicus*, *Squatina californica*	
Shellfish		
Mussels	*Mytilus californianus*, *Septifer bifurcatus*	Major food species
Sea urchins	*Strongylocentrotus purpuratus*, *S. franciscanus*	Major food species
Abalones	*Haliotis cracherodii*, *H. fulgens*, *H. rufescens*	Food, tools, ornaments
Limpets	e.g., *Lottia gigantea*, *Megathura crenulata*	Food, ornaments
Clams & Cockles	*Protothaca staminea*, *Tivela stultorum*, *Clinocardium nuttalli*	Food, ornaments
Turban & Top snails	e.g., *Tegula funebralis*, *Astraea undosa*, *Norrisia norrissi*	Food, tools
Crabs	*Brachyura* sp.	Food
Olive snails	*Olivella biplicata*, *O. baetica*	Ornaments
Cowry & Coffee bean	*Cyprea spadicea*, *Trivia californiana*	Ornaments
Tusk shell	*Dentalium pretiosum*	Ornaments

Notes: Does not include many minor or incidental midden constituents. Also recovered were human (*H. sapiens sapiens*), dog (*Canis familiaris*), and fox (*Urocyon littoralis*) remains—the canids probably introduced to the island by Native Americans—and the remains of an extinct mouse (*Peromyscus nesodytes*) and vampire bat (*Desmodus stocki*).
Data compiled from Rozaire (1976); Guthrie (1980); Walker (1980); Rick, Erlandson, and Vellanoweth (2001).

the seals had completely disappeared from the archaeological record of the North Island, and by 1790 their breeding range had shrunk to the southern part of the South Island. Historical exploitation by the European sealing industry further reduced the range and numbers of fur seals until protection was implemented in 1873. This is a fascinating example of the

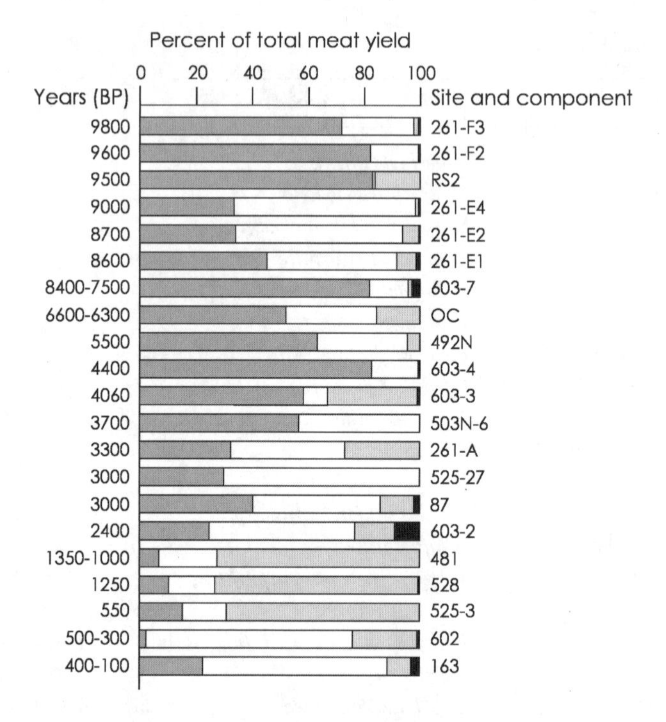

FIGURE 8.4. The archaeological record: prehistoric human diets. Reconstruction of marine resource use by prehistoric people at San Miguel Island, California, over the last 10,000 years. The data show percent of estimated total meat yield for shellfish (*dark gray*), fish (*white*), sea mammals (*light gray*), and birds (*black*) estimated from archaeological remains in shell middens. (Data adapted from Erlandson et al. [2005]).

initially strong depression of a marine resource by indigenous people and subsequent exploitation by Europeans, and the value of archaeological data for determining a baseline of geographic distribution. Many similar examples are emerging through collaborations of archaeologists and historical ecologists.

Analysis of stable isotopes and trace elements in human bones and teeth can be used to determine the health, nutrition, and diet of people in the past. Stable isotopes of carbon (C) in human bone collagen act as an indicator of past diets, because marine and terrestrial proteins leave different $^{13}C/^{12}C$ ratios. For example, inhabitants of Britain rapidly switched from a marine-based to a terrestrial-based diet following the introduction of agriculture at the onset of the Neolithic period about 5,200 years ago.

Archaeological remains can provide long time series of data spanning hundreds or thousands of years, with a resolution of decades to centuries

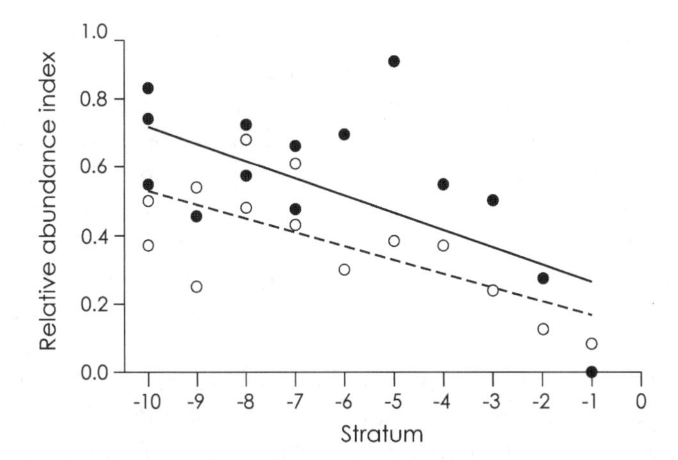

FIGURE 8.5. The archaeological record: resource depression. Reconstruction of resource use by prehistoric people. Archaeological remains from Emeryville Shellmound, San Francisco Bay, California, show trends of decreasing relative abundance of sturgeon (*solid circles, solid regression line*) and geese (*open circles, dotted regression line*) over time from the oldest (−10, ca. 2600 BP) to the youngest (−1, ca. 700 BP) stratum. (Data adapted, with permission, from Broughton [1997, 2002]).

(table 8.1). Data represent the local environment over a generally small to medium spatial scale. For much of the last 15,000 years, coastal shell middens may represent one of the best sources of information about the nature of past intertidal and near-shore communities, since rising postglacial sea levels have drowned, damaged, or destroyed much of the paleontological record for such ecosystems except in areas of active tectonic uplift. However, the record is biased to favor species that are well preserved, by soil and other geomorphic characteristics, and by stratigraphic mixing caused by burrowing animals, plowing, and other disturbances that limit temporal resolution. The quality and quantity of data also depend on the excavation and analytical techniques applied. Excavated remains provide valuable records of species' occurrences, size, and age. They are less valuable for estimates of relative abundance because the occurrence of remains is strongly biased by human activities, interests, and preferences (table 8.1).

Genetic Data

The genes in animals or plants living today, and ancient DNA extracted from remains of extinct species, represent an archive of information about species through deep ecological time. Levels of neutral genetic variation increase with population size. Thus, the level of genetic diversity found in a population today should reflect its average population size in the past. Genetic estimates of the average long-term population sizes of humpback, fin, and minke whales in the North Atlantic before intensive whaling suggest populations on the order of 240,000, 360,000 and 265,000 individuals, respectively. Palumbi discusses the underlying assumptions of genetic reconstruction of whale populations more fully in the next chapter. Genetic estimates of the minimum numbers of whales that must have been alive in more recent times are proving useful in evaluating the effects of whaling. In general, however, genetic data can provide range estimates of population sizes in the past, but these estimates tend to integrate population trends over large temporal and spatial scales (table 8.1). On the other hand, genetic data can provide reliable estimates of past population bottlenecks, which have proven to be very useful for tracking changes in the geographic distributions of species in response to past climatic change.

Historical Data

Historians use a variety of published, handwritten, and illustrated sources in libraries, archives, and museum collections to extract information about the past. Even before writing was invented, people left paintings and carvings in caves and on walls, and they left monuments of whale bones and other animal remains that can be used to trace their lives and activities. For example, a Neolithic Korean sandstone carving shows a whale harpooned by a boatload of men (figure 8.1C). It was discovered in 1971 and may be one of the oldest illustrations of whaling in existence. After the invention of writing, people left anecdotal descriptions of their activities, of the natural world around them, of their ordinary life, and of their travels and discoveries. They also left maps, drawings, and paintings. At some point in history, people started to leave records on trade and customs, catches of fish, the amounts of salt used to cure the catch, the technology of exploitation, and the regulations implemented as catches declined. These records became more detailed and sophisticated over time, eventually leading to modern fisheries and hunting statistics.

Historical sources provide information on species' occurrence, abundance, and distribution, as well as information about human impacts on species that were used for food, fuel, fashion, or other purposes. Richard Hoffmann used a wide range of written and illustrated information to reconstruct fisheries in medieval Europe. Freshwater and diadromous fish species originally preferred as food became scarce in the Middle Ages due to overexploitation, habitat degradation, and pollution. People responded to this depletion by implementing fisheries regulations, inventing aquaculture to raise carp, and expanding to saltwater fisheries. Similar human responses have occurred many times and in many places to this day.

In another example, old logbooks from whaling vessels were used to reconstruct former seasonal distributions of whales in the ocean (figure 8.6). These data suggest that the historic feeding range of humpback whales may have included portions of the mid-Atlantic ridge, which is very different from today's coastal migration routes to northern feeding grounds. More recently, Josephson and colleagues used similar data to correct our understanding of the historical distribution of right whales in the North Pacific.

Historical descriptions and records of fisheries can be used to reconstruct catches, effort, harvest rates, and, sometimes, relative abundance over time, as discussed in Bolster, Alexander, and Leavenworth in chapter 6 in this volume. Hutchings and Myers reconstructed historical catches and harvest rates of Atlantic cod in Newfoundland from 1500 to 1991, showing that the first signs of overfishing appeared in the mid-1800s. Historic and modern fisheries statistics also have been used to demonstrate the shift in fisheries from predominantly groundfish in the nineteenth and early twentieth centuries to invertebrates and plants in the late twentieth century, and an increase in numbers of target species in the commercial fisheries (figure 8.7).

Historical data are invaluable for understanding the scope of past fisheries and their decline, but they are biased by selective interest in different target species (table 8.1). Some of the better historical records can be used to estimate the relative, and even absolute, abundance of once commercially important species with reasonable confidence, as well as presence or absence of species. In general, historic data often have a high temporal resolution, but consistent records usually cover only short periods of time. The spatial scale ranges from local to global (table 8.1).

Early Scientific Data

In the nineteenth century, marine scientists started to systematically catalogue and investigate life in the sea. In some fortunate cases, regular measurements of temperature or precipitation extend back as far as a century. Biological records are generally more episodic and are usually limited to

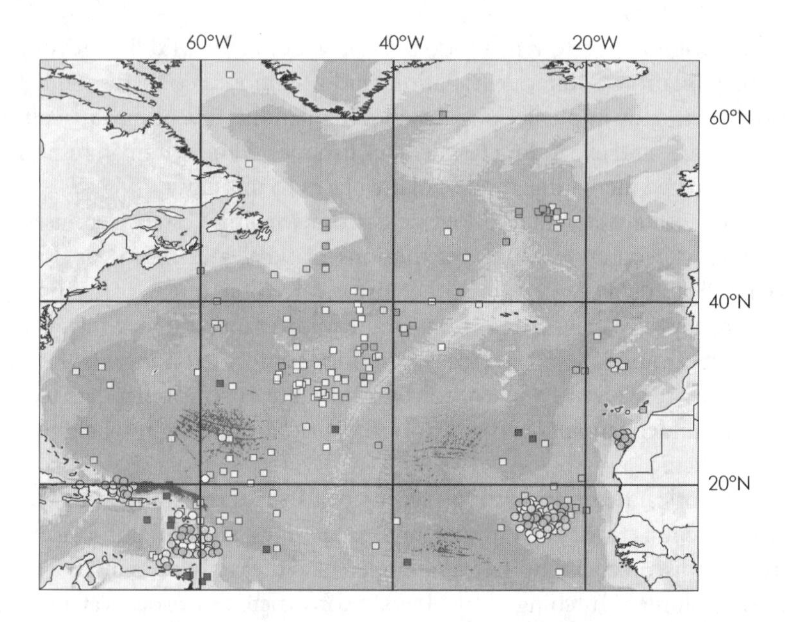

FIGURE 8.6. The historical record: whaling logbooks. Locations of humpback whales by month recorded in logbooks from whaling captains in the nineteenth century. Circles denote monthly catches extracted by Townsend in 1935, and squares denote sightings extracted by Reeves and colleagues in 2004. Records show a monthly progression from winter breeding grounds in the Caribbean and Cape Verde Islands areas, demonstrating a migratory distribution. The presence of humpback whales along the Mid-Atlantic ridge in the summer suggests historic summer feeding grounds that are far from northerly coastal feeding grounds known today (Smith et al. [1999]).

short-term investigations (table 8.1). Biologists published species lists and investigations on abundance, distribution, life history, and environmental parameters. In some cases, we can compare past with recent results if similar sampling effort or methods were applied. In other cases we can correct for differences in methods or adapt modern methods to simulate those used in the past.

Comparisons of early and modern studies for the Wadden Sea demonstrate extreme ecological degradation over the past century. Recent bottom surveys at the same sites investigated 50–100 years ago unambiguously demonstrated the loss of oyster banks and of complexity and diversity in benthic communities due to overfishing. Comparing phytoplankton samples from the 1930s and 1990s demonstrates changes in the abundance and frequency of blooms, as well as shifts in species composition from

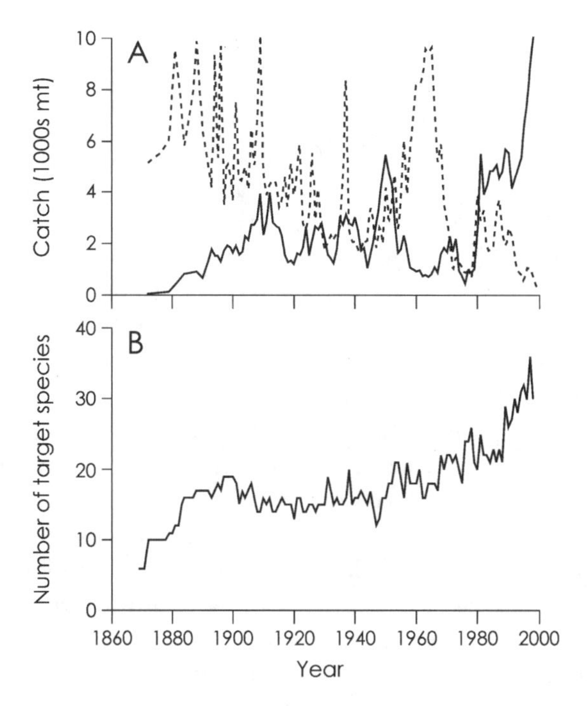

FIGURE 8.7. The historical record: fisheries statistics. Historical and modern fisheries statistics were used to demonstrate that (A) declining catches of traditional groundfish (cod, haddock, pollock; *dotted line*) are gradually replaced by increasing invertebrate and plant landings (*solid line*) in the Quoddy region of the Bay of Fundy. Note that high groundfish landings in the 1960s arose from expansion to offshore fishing grounds and the introduction of otter trawls. (B) Increase in the number of target species in commercial fisheries over time. (Data adapted from Lotze [2004]; Lotze and Milewski [2004]).

diatoms to dinoflagellates, a common sign of eutrophication. Likewise, recent surveys on the southeastern Australian shelf and slope reveal that changes in species composition began almost from the start of the trawl fishery a century before. None of these insights are possible with modern ecological data alone, although the time frame is generally limited to a century or less (table 8.1).

More systematic and detailed fisheries data have become increasingly available since the late nineteenth to early twentieth centuries. Despite their limitations, such data often provide the best long-term estimates of changes in abundance over time. A wide range of methods has been developed to analyze the data to infer past trends in abundance. Records of fishing effort and efficiency in addition to catch records can be used to calculate standardized catch rates as a measure of relative abundance over time. Estimates of present species abundance, recruitment, and natural mortality can be used to reconstruct former species abundance using simple population models such as Sequential Population Analysis. Data on stomach contents and diet can be used to calculate the former abundance of multiple species and their interactions using Multi-Species Virtual Population Analysis or other models. For example, Whitehead used a simple population model to estimate the trajectory of global sperm whale abundance from 1700 to 2000. Calculated abundance was 1.1 million before whaling began, compared to the 360,000 sperm whales that exist today. McClenachan and Cooper used a simple model to estimate that the population of now extinct Caribbean monk seals would have consumed a biomass of fish and invertebrates six times that found on typical reefs today.

Fisheries and hunting data are often the best and most consistent long-term, quantitative data available for the recent and historical past (table 8.2). However, because harvesting records were usually kept for commercial purposes and because harvesters adapted their searching and catch-handling behavior to local conditions, interpreting these records is difficult, especially for abundance. Powerful analytical methods are available for calculating past or future population size if we have good estimates of present abundance and life-history parameters. The greatest problems are that fisheries data are naturally biased toward species of greatest commercial interest and data can be skewed for political reasons.

Recent Surveys and Experiments

Increasingly, modern surveys and monitoring are providing consistent biological and environmental data. Monitoring includes a wide range of environmental parameters as well as the occurrence, abundance, distribution, size, or health of selected species. Standardized annual surveys of fish abundance on continental shelves can be used to analyze population trends, spatial distribution, and productivity of fish stocks over time. Spatial surveys, which compare sites with different abundance levels and sizes of predatory fish, can be used to infer how large predators affect community structure. Ecological surveys of protected and unprotected areas can be used to quantify the effects of human exploitation on the abundance and size of intertidal invertebrates.

Using a more historical approach, Roy and colleagues used information from museum collections in conjunction with field surveys to show that human activities have led to significant and widespread declines in body sizes of rocky intertidal gastropods since the 1960s in southern California (figure 8.8). These declines are not restricted to species harvested for human consumption such as the gilded tegula (*Tegula aureotincta*) and the owl limpet (*Lottia gigantea*). For example, the volcano keyhole limpet (*Fissurella volcano*), which is not commonly harvested for food, is declining due to accidental catch and chronic human disturbance. Comparable declines did not occur at a nearby protected area where conservation laws are strictly enforced (figure 8.8). Similar changes in response to human harvesting have been documented for many other species.

Standardized surveys and monitoring provide detailed trends of biological or environmental changes that rarely extend back more than fifty years (table 8.1). Thus, apparent trends may reflect phases of longer cycles, or even random fluctuations in ecosystems that were greatly altered by human disturbance before monitoring began. However, the temporal and spatial resolution of monitoring and survey data is very high, and we can obtain relative and sometimes absolute estimates of abundance, density, or other measures. This greater precision helps to tease apart cause and effect, although data can be strongly biased by the purpose and design of the surveys, methodology, and effort.

Experiments help ecologists to ask questions and test hypotheses about factors that determine the distribution and abundance of species, communities, or other elements of nature. Experiments are a powerful tool to tease apart the causes and consequences of ecological changes and their magnitude, as well as the occurrence of interactions among species, and the nature and magnitude of human impacts. For example, both nutrient supply and consumer pressure influence the structure, diversity, and functioning of plant-dominated benthic communities. However, the magnitude and direction of these "bottom up" and "top down" effects depend on each other and on the productivity of the system. Nutrient supply and consumer pressure also interact with environmental factors such as temperature and ultraviolet radiation to affect germination and growth of macroalgae.

Such experiments demonstrate that it is rarely a single variable that determines species performance, community structure, or ecosystem function, but that multiple environmental and ecological factors interplay. They further suggest that multiple human impacts such as eutrophication, alterations of food webs, and climate change have interdependent effects on ecosystems. Thus, a comprehensive understanding of the current state of an ecosystem requires knowing the historical changes in the environment, in species, and in human activities of all kinds.

Not all experiments are planned in advance. Some occur by chance.

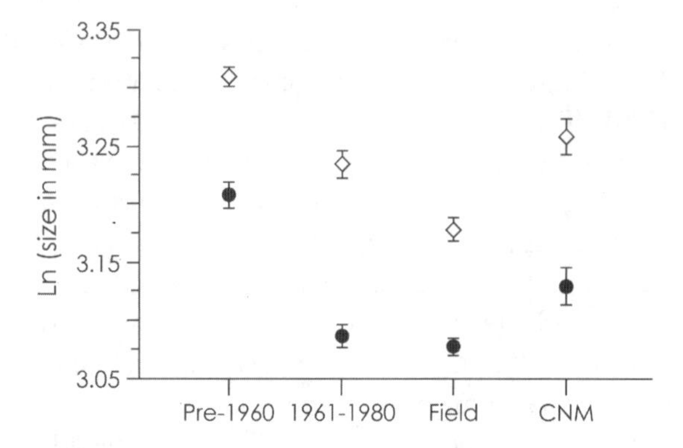

FIGURE 8.8. Survey data. Average size of two intertidal gastropods, the gilded limpet tegula, *Tegula aureotincta* (*open diamonds*) and the volcano keyhole limpet, *Fissurella volcano* (*solid circles*) at the southern California coast at two periods in the past (data from museum collections), from a recent field survey, and at the protected area Cabrillo National Monument (CNM) for comparison of recent exploitation effects. Data are means and 95 percent confidence intervals. (Adapted from Roy et al. [2003]).

Trawling for fish was greatly reduced during World Wars I and II. These interregna resulted in much greater catches of larger fish immediately after cessation of hostilities, although the benefits were short lived due to overfishing during postwar fishing booms. Nevertheless, experiments are the most rigorous way to evaluate causal relationships and elucidate mechanisms of change. Results inevitably depend on the experimental design, and the appropriateness and scale of experimental conditions, which may range from laboratory beakers and aquaria, to larger outdoor tanks, to experiments at sea (table 8.1). In general, the better the experimental control, the less the experiments resemble natural conditions. So-called natural experiments such as the effects of the two world wars on fish stocks are invaluable, but the outcomes are not always so clear. Combinations of experiments and research surveys may be the best option to test hypotheses and to validate their relevance in nature, but experiments can only suggest and never prove causes of past changes.

Integrating Data from Different Temporal and Spatial Scales

Historical ecologists attempt to reconstruct the abundance, distribution patterns, and food web links of individual species in the past, as well as the structure and function of the ecosystems in which they lived, in order to determine trajectories of change into the present. Our goal is to analyze the direction, magnitude, and rate of change, as well as its causes and consequences. To this end, we can view the wide range of data, methods, and disciplines as pieces of a puzzle. The puzzle and its pieces can be qualitative, conceptual images of the past, or more quantitative reconstructions. Especially for single species, it is sometimes possible to reconstruct past population abundance and distribution with considerable confidence. In most cases, however, we can only estimate relative abundance in comparison to other species living at the same time in the past or to the abundance of the same species living today. Inevitable gaps in the puzzle can be filled in using mathematical models to estimate missing values.

The three most successful approaches to reconstructing the past and analyzing patterns of ecological change include comparing different times in the past with the present (then versus now), analyzing time series of ecological change, and reconstructing ecosystem models of past food webs and communities.

Then Versus Now

Comparing species lists from different times in the past provides a simple but powerful means of detecting past extinctions, introductions, or dramatic changes in abundance. Wolff used archaeological records, historical descriptions, and recent distribution patterns to reconstruct the former fauna and flora of the Wadden Sea and the extinctions that occurred over the past two thousand years due to large-scale human exploitation and habitat alteration. Likewise, Lotze and Milewski estimated that more than half the marine bird and mammal species in the outer Bay of Fundy were severely reduced or extinct by 1900 due to human impacts. We can also estimate former abundance based on the amount of habitat that was available in the past, as has been demonstrated for waterbirds by using old maps of the Netherlands. Similarly, estimates of the loss of spawning habitat for Atlantic salmon and gaspereau (alewife, *Alosa pseudoharengus*, and blueback herring, *A. aestivalis*) in the St. Croix River, New Brunswick, due to large-scale damming of rivers and pollution beginning in the 1800s, agree well with historic catches described in fisheries reports from the nineteenth century. These were more than a hundred times greater than today (figure 8.9).

Some of the greatest losses observed are for large fish species. In the Quoddy region of the Bay of Fundy, Canada, catches of haddock (*Melanogrammus aeglefinus*), halibut (*Hippoglossus hippoglossus*), pollock (*Pollachius virens*), and cod in the 1990s were only 3 to 37 percent as great as catches a century before, despite great increases in effort and the efficiency and spatial extent of fisheries (figure 8.9B). Likewise, standardized catch rates of the most commonly caught species of sharks in the Gulf of Mexico have declined 90 to 99 percent between the 1950s and 1990s, and the same is true for large predatory fish communities from nine different study areas worldwide. These and other historical estimates suggest that exploitation has fundamentally altered marine food webs and ecosystems over time.

Time Series

Repeated historical observations of absolute or relative abundance can be analyzed mathematically to calculate trends back in time and into the future. Different time series of environmental and biological data can be correlated to look for interactions between species and environmental parameters, or between different species, and to infer causes of change over different spatial scales.

Pandolfi and colleagues used paleontological, archaeological, historical, and recent scientific data from coral reefs around the world to estimate the categorical abundance of different groups of reef organisms over different cultural periods. Overall trajectories of ecosystem decline were strikingly similar worldwide despite strong regional differences in types of reefs and the timing of the onset of intensive degradation. Similarly, worldwide patterns of decline have been documented for estuaries and coastal seas. An example for the outer Bay of Fundy is shown in figure 8.10. In general, marine mammals and birds declined before fishes, invertebrates, seagrasses, and wetlands. Hunting, fishing, and habitat loss reduced animal populations, dike building and draining destroyed marshes, and seagrass succumbed to wasting disease. Only birds have shown any significant evidence of recovery.

Time series of standardized fisheries and monitoring data are especially powerful for estimating changes in fish populations, teasing apart the relative contributions of fishing versus oceanographic change, and understanding the effects of fishing on food webs. Unfortunately, such time series rarely extend back more than a few decades. Records thousands of years into the past can be obtained by analyzing fish scales or otoliths in laminated marine sediments, as has now been done for fish stocks offshore in California and South Africa. We can also use changes in the relative abundance of microfossil species to infer changes in environmental condi-

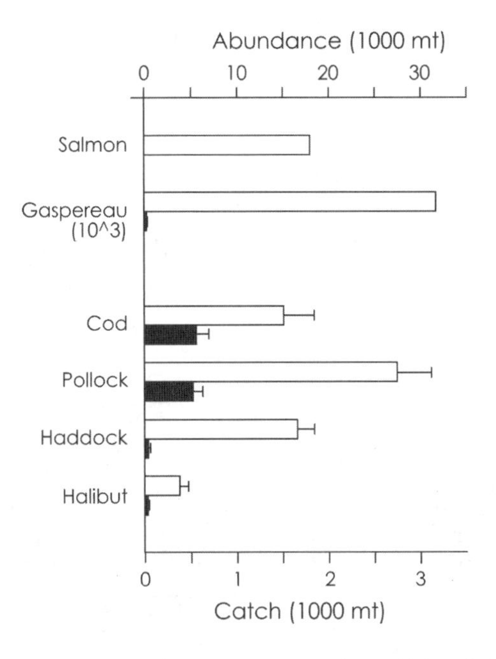

FIGURE 8.9. Historical comparisons. Reconstruction of historical (*white bars*) and comparison with recent (*black bars*) abundance or catch of important fish in the outer Bay of Fundy, Canada. *Upper scale*, Reconstructed pre-1800s abundance of Atlantic salmon and gaspereau (alewives and blueback herring) in St. Croix River based on historic fisheries descriptions and modern estimates of available spawning habitat, and recent counts of returning fish. *Lower scale*, Comparison of average historic (1890–1900) with average recent (1990–2000) groundfish catch in the Quoddy Region. (Data adapted from Lotze and Milewski [2004]).

tions for which correlations with fish stocks or other environmental variables can be established.

Food Webs

Historical data can be used to reconstruct food webs based on the presence or absence of different taxa over time. In addition, basic information on their diets can be compiled mostly from stomach contents or by comparisons with other, better studied species whose diets are inferred to be similar. So far, results are similar regardless of the types of qualitative or quantitative data employed or the methods of analysis. Large apex predators are disproportionately removed first, often resulting in the phenomenon of "fishing down the food web" as well as overall dramatic losses in biomass.

Toward a Vision of the Past

Historical ecological analysis is imperfect and fraught with difficulties of incomplete and approximate information. Nevertheless, careful analysis

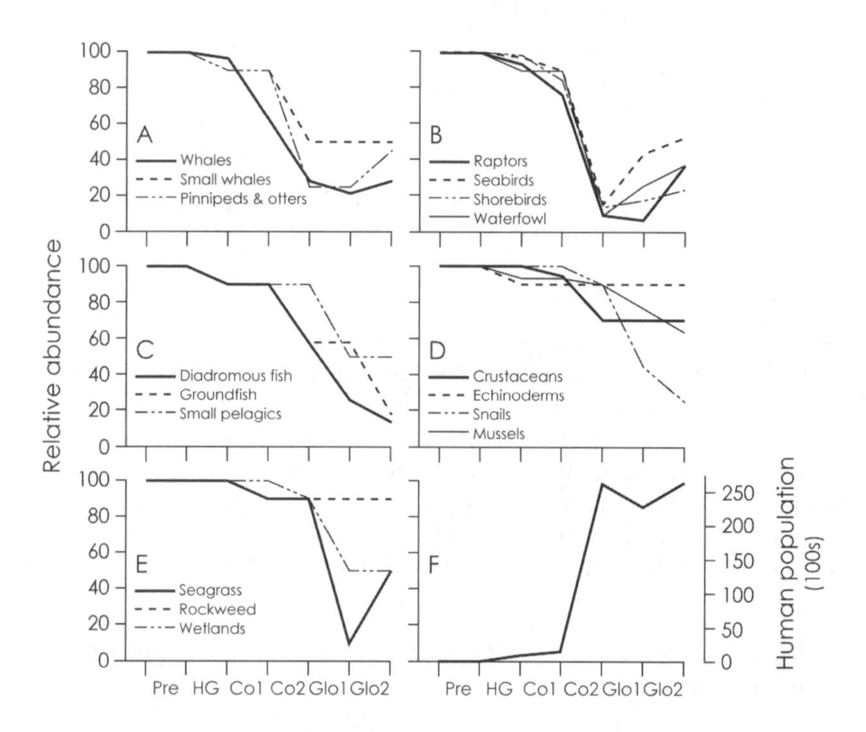

FIGURE 8.10. Integration of different disciplines to create long-term timelines of change. Relative abundance of (A) marine mammals, (B) birds, (C) fish, (D) invertebrates, (E) vegetation, and (F) human population in the outer Bay of Fundy, Canada, over different cultural periods in the past. Relative abundance was estimated in percent of original levels or in categorical abundance as pristine (100%), abundant (90%), depleted (50%), rare (10%), and extinct (0%). Cultural periods were, *from left to right*, Pre (pre-human: >8000 BC), HG (hunter-gatherer: 8000 BC–AD 1600), Co1 (colonial establishment: 1600–1760), Co2 (colonial development: 1760–1900), Glo1 (early global: 1900–1950), Glo2 (late global: 1950–2000). (Data adapted from Lotze and Milewski [2004]; methods after Lotze et al. [2006]).

demonstrates that humans have affected the marine environment ever since they began to collect shellfish for food and that these impacts have intensified in variety and magnitude of disturbance, and at ever increasing rates in recent centuries. The most obvious changes include decreases in species abundance and size, and changes in community composition, which alter food web structure and function to degrees we are only beginning to understand.

Historical ecology also inspires our imagination. Reading descriptions of how the world appeared centuries ago, sorting through reports of archaeological investigations, and measuring changes in organisms and environmental proxies from the sedimentary record help us to move beyond the data to develop an image of what the oceans were like before large-scale human alteration. Together, knowledge and imagination give us back what is lost in living memory: a long-term vision of the history of nature and of ourselves. This vision can give us a sense of origin, knowledge of the path we've traveled, and an idea of the magnitude of change that has taken place. We can use that vision to question our actions and their consequences for the ocean and to rethink and redirect our path into the future.

Acknowledgments

We are grateful to the late Ransom A. Myers and to Boris Worm for helpful discussions and comments. Financial support was provided by the Census of Marine Life's History (HMAP) and Future (FMAP) of Marine Animal Populations programs funded by the Alfred P. Sloan Foundation.

Chapter 9

Whales, Logbooks, and DNA

STEPHEN R. PALUMBI

Historical Data Need Careful Analysis

Sixteenth-century Venice was a fascinating place. The *glitterati* of the Italian Renaissance were close by. The *canali* (the sewage system) were the best in Europe. The economy was booming. Although comings and goings of the wealthy and powerful are well recorded almost everywhere, the history of common commerce is particularly clear in Venice, a center of the European merchant class. "I can without exaggeration claim to see the dealers, merchants and traders on the Rialto of the Venice of 1530 . . . ," assures Fernand Braudel in his extensive economic history of Europe. This claim is abundantly documented in a superb written record of the ins and outs of Venetian market commerce—but there is a danger in using these data to answer questions for which they were not designed.

Take, for example, the evidence for economic stability in late-sixteenth-century Venice. Detailed price records show that the cost of a loaf of bread in Venetian markets varied by less than 20 percent during the last fifteen years of the century. Because bread was the main market commodity for poorer workers, its price substantially dictated the quality of life for working people. Long-term stability in bread prices suggests long-term economic health and stability for all strata of society. However, this was a time

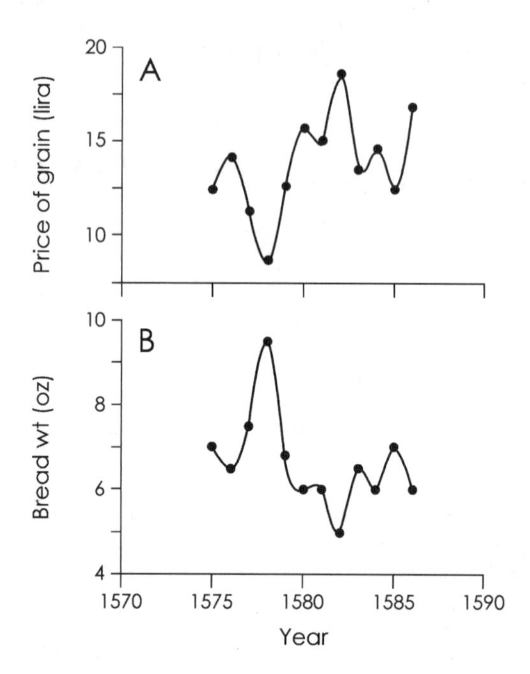

FIGURE 9.1. Grain prices and weight of bread loaves in sixteenth-century Venice. (A) Grain prices (lira) varied twofold in Venice at the end of the sixteenth century, despite stable prices for bread. (B) The weight of bread loaves sold in late-sixteenth-century Venice. Bread prices could remain stable in the face of varying grain prices because the size of a loaf of bread changed with grain cost. (Data from Chapter 2 in Braudel [1978]).

of upheaval, and other data show that grain prices changed wildly during this time (figure 9.1). If grain prices varied, but bread prices did not, one possibility is that Venice, like ancient Rome, sold grain below market price in order keep the cost of bread artificially low in times of economic turmoil.

This historical conclusion may appear logical and compelling, but it turns out to be wrong. It depends upon an inadvertent but critical assumption about the historical record. In fact, although the price of a loaf of bread was stable, Venetian bakers varied the *size* of loaves depending on the price of grain (figure 9.1), a practice that was common in Renaissance Europe. As a result, a loaf was always the same price, but in hard times, that loaf would be up to 50 percent smaller and feed far fewer people. Stable bread *prices* do not provide conclusive evidence of economic stability.

Braudel's data are not at fault in the above example. But the use of historical observations to answer questions for which they were not intended presents an inherent problem—the need for accurate contextual information. If we supply context from our own experience—for instance, if we assume that loaf sizes are constant and prices vary because that is what we find in today's supermarket—we risk misreading the historical record because this was not the reality of hungry Venetians in the late 1500s. Similarly, interpreting historical data about state of the oceans in the past requires making assumptions about the past. In many cases, the assumptions underlying data analysis may be more important than the data themselves in determining the sensitivity of the analysis and the likelihood of the outcome.

Sensitivity about Whales

Alternative assumptions affecting the interpretation of historical data are clearly important in reconstructing the size of whale population before whaling. This has practical importance because, at its inception, the International Whaling Commission received a global mandate to monitor the current state and determine the past condition of whale populations in order to manage hunting in the future. However, the size of whale populations in the past is difficult to estimate. One approach has been to model populations in terms of removals and to use historical catch records to estimate how many whales there must have been before removals began. This process mines historical data and filters them through a set of assumptions about population growth in order to generate the results.

Various methods have been used to model this process, but the basic approach has been, for each whale species, to estimate the population growth rate, subtract the known catches, and determine what pre-whaling population size accounts for the current numbers, which are known from careful population surveys. The simplest versions of these equations, known as BALEEN II, are independent of age and sex. They relate population size in year t (N_t) to population size in year $t - 1$, maximum possible growth rate (r_{max}), the population level at which a fishery produces maximum sustainable yield (μ), the overall environmental carrying capacity (K), and hunting mortality, that is, the number of whales killed by hunting, in year $t - 1$ (H_{t-1}):

$$N_t = N_{t-1} + r_{max} N_t[1 - (N_t - 1/K)^{\mu}] - H_{t-1} \quad (1).$$

Since hunting mortality is the only variable dependent on whaling, historical whaling records play only a small role in determining the population trajectory. Nevertheless, several sets of crucial assumptions must be applied in order for the historical data to be useful in equation 1. First, although the equation requires knowing the hunting mortality, hunting mortality is not known for any whale population. Instead, it must be estimated from catch records. Here, historical data are often limited to numbers of barrels of oil that documents show were off-loaded by ships or, less frequently, to the number of whales of different species rendered. The best data come from capture records, but even these estimate only that fraction of all the whales killed that were recorded by the whalers and tried out for oil. Even with perfect records from individual whalers, the total capture for a whole population would require knowing all the records of all the whalers. To estimate hunting mortality from incomplete catch records, we must estimate the fraction of missing historical records and estimate the number of whales killed, but lost to the fishery. As a result, hunting mortality in year t (H_t) is related to the total recorded catch that year (C_t) by

$$H_t = C_t{*}(1/R){*}[1/(1-L)] \qquad (2),$$

where R is the fraction of the original catch that is currently compiled in extant historical records and L is the fraction of whales killed, but lost at sea. Other adjustments might also be required.

Three other sets of assumptions must be added to equation 1. The variable r_{max} is the maximum rate at which a whale population can expand, and, in this equation, it must account for all births and all deaths except for those due to hunting. The term r_{max} is familiar to ecological demographers. It refers to the maximum innate capacity of a population to increase and is highest for populations far below carrying capacity. Measuring r_{max} is difficult and demands observations of population growth when the population is small. The use of r_{max} in equation 1 also assumes no interaction among species—growth rate is fixed and independent of the abundance of competitors or predators, or the availability of food. In this, the definition of r_{max} departs significantly from the definition typical in ecological demography.

Variation in population growth over time derives from two other parts of the equation—the carrying capacity (K) and the population level that supports maximum sustainable yield (μ). As used here, K is the population size at which the growth rate is zero, and it necessarily depends on the same parameters as r_{max}. The factor μ is the slope of the curve relating yield to the standing stock and is typically envisioned as that fraction of the carrying capacity at which the population is at maximum sustainable yield (MSY). For example, if the population reaches MSY at 50 percent of carrying capacity, then μ is 0.50.

Rigorous historic reconstructions can be accomplished only if r_{max}, μ, total recorded catch (C), completeness of historical records (R), and the rate of whales killed but lost at sea (L) are known. Then, estimates of possible values for K can be made by varying K and testing which values allow the subsequent population trajectory to confirm historically known population sizes. Whitehead attempted to reconstruct the history of sperm whale populations in this way and showed that wide bounds, from 500,000 to 1,500,000, bracketed the number of sperm whales before whaling. Attempts to model the history of North Atlantic humpback whales have not been fully successful; model trajectories do not pass through well-known census values from recent decades. Gray whale populations in the eastern North Pacific are difficult to model demographically—known population parameters and the recorded hunting record do not combine to give a good fit of gray whale population trajectories to population counts in the twentieth century unless complex models with changing carrying capacities are invoked.

Uncertainty in Reconstructing Historical Whaling

Because of the difficulty in using historical data to model past whale populations and the complexity of some of the models that are used to bring historical and current population data into alignment, it is worthwhile to explore the sensitivity of historical reconstructions to simple variation in assumptions about R and L. Although it is beyond the purposes of this review to estimate either variable, previous studies have used a range of values for L, the proportion of whales killed, but lost at sea. Edward Mitchell and Randall Reeves suggested a loss rate of L = 50 percent ($1/[1-L] = 2.0$) in preindustrial whaling in the North Atlantic. Peter B. Best and colleagues suggested L = 25 percent ($1/[1-L] = 1.3$) for right whales, but J. E. Scarff suggested 57 percent ($1/[1-L] = 2.3$). Stuart C. Sherman notes that whalers typically lowered boats to initiate a chase five to six times for each whale caught, but the fate of the many whales that were chased and lost was seldom recorded.

Quantitatively, the completeness of historical records (R) is virtually unexplored for nineteenth-century whaling. Of 13,927 whaling voyages tallied by Sherman, only 4,000 logbooks were accessible to researchers at that time, or 30 percent. It seems reasonable to assume that records of industrial whaling would have become more accurate over time and that documentation for later voyages would be more complete. However, countervailing trends raise questions about these assumptions. In particular, Reeves noted that the quality of logbook entries often declined from the 1840s to the 1880s, providing less information per hunt or per voyage.

These difficulties are surmountable with extra effort paid to the historical data, the assumptions used to interpret the data, and the final analysis. Figure 9.2 shows how total hunting mortality is dependent upon assumptions about the completeness of historical data, described in equation 2. Given a recorded humpback whale catch in the North Atlantic of about 29,000 animals, hunting mortality can be estimated at various levels of hunting loss (L) and for various assumptions about whaling record completeness (R). If L and R are both 50 percent, for example, then total hunting mortality is about 120,000. This figure is a simplification: loss rates probably vary decade by decade and species by species. Sherman suggests that "interpretation of the extracted information by scholars is essential," and figure 9.2 shows why careful documentation of the analytical assumptions is just as important as careful extraction of the raw whaling data.

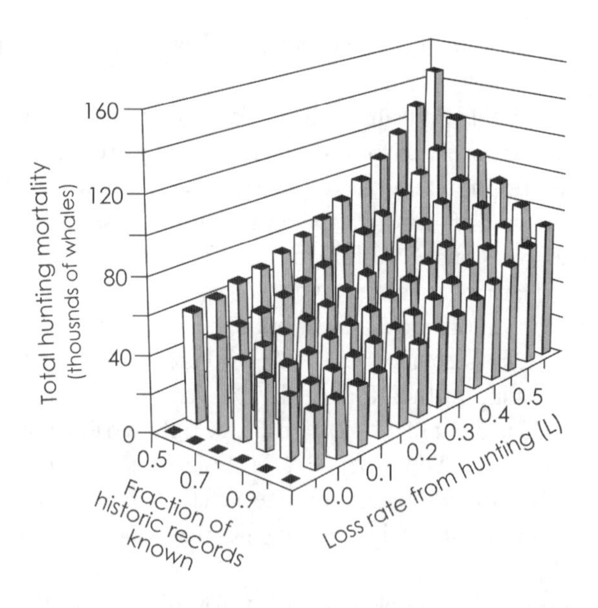

FIGURE 9.2. Estimates of total hunting mortality for humpback whales in the North Atlantic depends strongly on the fraction of historic records currently accounted for and the loss rate from hunting. The sensitivity of hunting mortality is based on compilations of catch records for humpback whales in the North Atlantic from 1650 to 1910 and varied assumptions about possible values for historical completeness and hunting loss.

A Genetic View of the Past

Historical data analysis and interpretation benefit from comparison with other methodological approaches, which use independent data sets to address the same question. Recently, genetic data have been explored as a means to illuminate the past population sizes of hunted whale populations. Genetic diversity within a population builds up due to mutation and is reduced by genetic drift, which happens due to inbreeding. Small populations subject to severe inbreeding have a lower amount of genetic diversity. By contrast, larger populations with little inbreeding show a higher amount of diversity. Genetic diversity in whales can be measured by comparing DNA sequences that have been obtained from individuals in the same population. Since the movement of whales from population to population can inject genetic variation into the group, migration needs to be taken into account. Migration can be measured by comparing the DNA sequences of individuals from different populations. As a result, analyzing DNA sequences from multiple populations of whales can help estimate migration patterns and measure levels of genetic variation. If DNA sequences are only affected by drift and mutation, then sequence diversity (represented by θ) is proportional to the product of the mutation rate (μ) and the effective population size. For nuclear DNA, inherited through both the male and the female lineages, this relationship is denoted by

$$\theta = 4N_e \mu \qquad (3),$$

where N_e is the long-term effective size of the population.

Genetic variation is related not just to the current size of whale populations, but to population levels in the distant past. This is because mutation builds up diversity slowly, and drift removes variation slowly. For a region of whale DNA in an intervening, noncoding sequence in a coding gene, mutation adds about 0.5 percent to sequence diversity in 100 million years. Genetic drift due to inbreeding removes some of this diversity every generation, and the fraction removed can be estimated as $1/4N_e$. However, this removal is also slow. For North Atlantic right whales, the most threatened whale species known, breeding adults probably number about 100. Even with this very low number, drift removes only 1/400th of the current nuclear genetic variation in right whale DNA each generation. As a result, current levels of whale diversity reflect past population sizes more than they reflect current levels.

Genetic diversity appears to be very high when current estimates of past population sizes are taken into account. For example, the International Whaling Commission estimates that humpback whales around the world numbered no more than 115,000 before whaling. Total genetic diversity (θ) in humpback whale mitochondrial D-loop sequences is about 10 percent worldwide. The mutation rate (μ) has been measured to be about 2–4 percent per million years, or 20–40 percent per million generations, in this region of the whale genome. Using these data, the effective population size of humpback females worldwide is estimated to have been $\theta/2\mu$ = 125,000–250,000. Mature breeding females are thought to make up about one-sixth to one-eighth of a whale population, so these numbers suggest a global humpback whale population size of about 750,000 to 2,000,000 animals. Genetic estimates for gray whales, based on multiple nuclear loci, suggest a North Pacific population of about 90,000 individuals instead of the 20,000–30,000 typically assumed in models and management plans.

Like population sizes reconstructed from historical logbook data, genetic reconstructions also depend on a set of quantitative assumptions. Mutation rate, generation time, and the ratio of breeding to nonbreeding adult females are particularly critical. For example, if mutation rates are an order of magnitude higher than estimated, then population size estimates would be an order of magnitude lower. If all adults bred equally well and there were no animals that were nonreproductive as adults, then total population estimates would drop by about a factor of two. However, no survey of a wild mammal population has ever observed such extreme constancy in offspring production among parents. In all monitored populations, there is variation in reproductive success from adult to adult. In addition, the fraction of successfully breeding female whales is thought to be much smaller than the 50 percent assumed here.

It is possible that a combination of genetic parameters could give rise to population estimates wholly inline with previous thinking. However, in order for genetic models to reproduce Atlantic humpback whale population sizes before whaling of 10,000–20,000 similar to those estimated by the IWC,

- the mutation rate must be higher by a factor of two, *and*
- all females must breed at the same rate, *and*
- generation times in the past must have been 40–50 years, instead of the 20–25 years observed today.

In fact, our recent work suggests that the first condition is probably true: previous estimates of the mutation rate of one gene were low by about a factor of two. However, the mutation rates of other genes seem robust, and variation in the reproductive rate among females appears to be measurable. Perhaps future research will show these other assumptions to be faulty (the most fascinating assumption is the one on generation time: suppose whales lived a LOT longer in the past than we think they do today?). But until then, it is most probable that the signal of high genetic variation reflects a past in which Atlantic humpback and Pacific gray whales were more abundant than previously imagined.

DNA Provides a Long-Term Estimate

When did these whales live? The simple answer—before whaling—belies the complexity of genetic reconstructions. In fact, the historical populations chronicled in the genetic code could have lived long before industrial whaling began. Population sizes reconstructed genetically are typical of the species over long periods of time, and such long-term averages could include periods when numbers were low. However, if populations are low for long periods of time, genetic variation will be stripped away. For example, a population of 100 females loses 0.5 percent of its genetic variation every generation, or 40 percent of its variation over 100 generations.

Estimating how much variation is stripped away each generation by low population sizes places general bounds on the variation in size that could have occurred in the past. For example, genetic diversity in one population of North Pacific gray whales suggests a historic population of 90,000 instead of 20,000. Suppose this population was at some unknown but high level during the last glaciation, 18,000 years ago, and subsequently fell to 20,000 animals. If this were true, then the whale population would have lost about 1/5,000th of its mtDNA variation every generation since the reduction because only about 2,500 breeding females would have been left. Because 1,200 generations of humpback whales have occurred since the last glaciation, this population would have retained $(4{,}999/5{,}000)^{1{,}200}$ of its original variation, or about 78 percent. As a result, its size would have to have been about 22 percent higher than the previous estimate of 90,000 individuals in order to have retained the variation we see now.

A separate type of calculation can help us ask whether high diversity in a current whale population is likely to be due to a large population far in the past followed by a population crash well before whaling. Alter and colleagues simulated the diversity of DNA sequences in gray whale populations as if they had been subjected to a bottleneck during whaling and to a bottleneck some time farther back in the past. Comparing the genetic diversity of simulated populations with observed genetic diversity allows an estimation of which bottleneck timing is consistent with current levels of genetic variation. The result showed only a 5 percent chance that a population suffering a pre-whaling bottleneck 30 generations ago (300–450 years) would still retain the high level of variation seen in gray whales today. Bottlenecks farther back than that have even a lower likelihood (figure 9.3). As a result, it is unlikely that the high level of current gray whale genetic diversity was due to a large population far in the past that crashed before whaling began.

These calculations are little more than a cartoon of the way population trajectories affect genetic diversity, and they deserve continued modeling, but together with the genetic data on humpback, fin, minke, and gray whales, they suggest four strong conclusions:

- Any decline in whale numbers has been recent—within the past 100,000 years and probably since the last glaciation.
- The decline affected multiple species (fin whales, humpback whales, and gray whales at least).
- The species less hunted, minke whales in the Atlantic and the

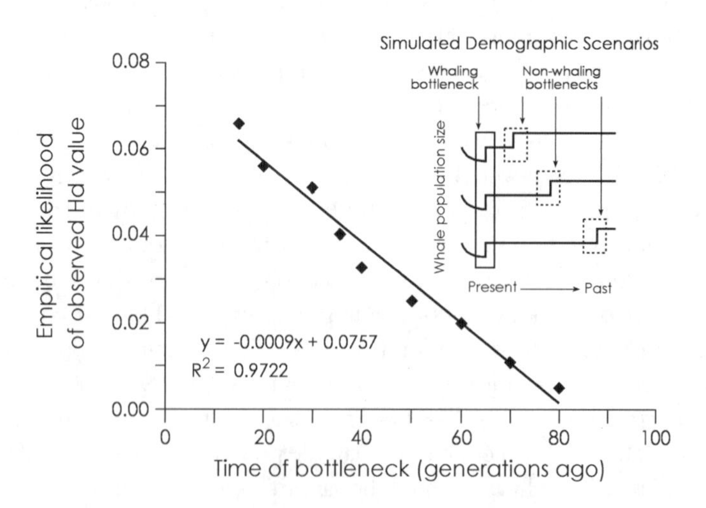

FIGURE 9.3. The likelihood of the observed genetic variation (Hd, haplotype diversity) in gray whales given a simulated bottleneck scenario beginning 15–80 generations ago (see inset drawing). A linear regression was used to determine the generation at which the likelihood falls below 0.05 (30 generations ago, or approximately 300–450 years). (Data from Alter et al. [2007]).

Antarctic, shows much less discrepancy between genetic estimates and current estimates of population size.
- The decline was worldwide.

We know of one factor that accounts for these patterns—historical whaling. Other factors may also have been responsible, but to date there has been no concerted attempt to provide reasonable demographic or historic alternatives. Genetic possibilities include a dramatic underestimate of the generation time of whales, a large increase in genetic mutation rate in current populations compared to estimates derived from comparisons among species, and a currently unknown demographic pattern that preserves genetic diversity.

Seeing the Past through Complex Lenses

Data from the past—whether they derive from DNA sequences or whaling logbooks—require a set of careful assumptions to convert them into useful conclusions. I have tried to point out ways in which data sets from genetics and logbooks might be carefully interpreted to provide a framework for the reconstruction of past whale populations. A challenge will be to improve the accuracy of assumptions on a wide variety of interrelated topics such as mutation rate, variation in female reproductive success and mutation processes in DNA, catch record completeness, hunting loss rates, and logbook accuracy. Though history and genetics seem like very different topics, they intersect in models that convert historical data into historical perspective. Attempts to improve this perspective are central to understanding the nature of marine ecosystems before they were fundamentally perturbed by humans and to charting the future of these ecosystems in a world wracked with human-induced change.

PART V

From Fisheries Management to Ecosystems

This book is about the history of fish and people and how understanding that history might help preserve the future of fish and the well-being of people. Fish will not become fruitful and replenish marine ecosystems on demand, nor will pollution, rising temperatures, or other physical insults abate simply because we want them to go away. To restore the ocean, people must be convinced to change their behavior because no policy will work without citizen resolve.

Fisheries science is necessarily geared toward fishing industry, management, and policy. As in other economic sectors, management has unwittingly encouraged cycles of boom and bust. New technology and the creation of new markets for alternative fish species temporarily restored prosperity, but each success reset institutional memory. This final section examines the role of historical perspective in governance and evaluates the practical benefit of the lessons learned.

In chapter 10, we return to New England. Andy Rosenberg is a fisheries scientist and experienced fisheries manager with ten years in the trenches of the northeast region of the National Marine Fisheries Service (NMFS), two as regional administrator in Gloucester, Massachusetts, and two more years as deputy director in Silver Spring, Maryland. Since leaving the National Oceanic and Atmospheric Administration, Rosenberg has written extensively on management and policy issues, served on the U.S. Oceans Commission, and advised the National Academy of Sciences and the United Nations on oceans policy issues. In the 1990s, he was at ground zero of the worst fisheries crisis in American history. His strong leadership reversed steep declines in a number of marine and anadromous species and set the bar for effective, though controversial, management.

In 1994, Rosenberg closed New England ground fisheries in order to rebuild cod, haddock, yellowtail flounder, and other stocks decimated by overfishing, which had accelerated in the 1980s. Fishermen were incensed, and, in 2002, some of them hung Rosenberg in effigy in Gloucester Harbor, protesting a new round of court-imposed restrictions aimed at rebuilding groundfish stocks. However, today haddock and sea scallops once again support valuable fisheries, and some other fish stocks are recovering. A few fishermen have even admitted privately that Rosenberg was right after all.

In 1999, Rosenberg also supported removing the Edwards Dam on Maine's Kennebec River to restore spawning habitat for salmon, sturgeon, alewife, and other anadromous fish. After ten years the lower Kennebec showcases the local benefits of dam removal. Alewives now return in the millions. Ospreys, eagles, and seals compete with sport fish for the bounty, and boating, fishing, and other recreational activities thrive, supporting local enterprises.

Karen Alexander shows that fisheries management is not new. Regulations that look remarkably familiar date back to the Middle Ages in Europe and start in 1623 in New England. Jamie Cournane currently serves on the Atlantic Herring Plan Development Team of the New England Fishery Management Council. She has worked most recently on mapping hot spots of river herring bycatch by large trawlers and seiners that target Atlantic herring. She discusses the success stories and explains how historical perspectives are increasingly important in management. Chapter 10 shows how understanding of shifting baselines can help to restore the oceans—if we stay the course.

In chapter 11, Enric Sala and Jeremy Jackson present lessons learned in the warmer waters of the Mediterranean and the Caribbean. Rather than return to the broad questions raised by Carl Safina in chapter 1 and debated throughout this book—How should history influence marine science, how should science influence marine policy, and how do science and history leverage each other to greater effect?—Sala and Jackson conclude with a new set of criteria to determine whether or not historical marine ecology may eventually fulfill its promise.

Chapter 10

Management in the Gulf of Maine

Andrew A. Rosenberg, Karen E. Alexander, and Jamie M. Cournane

A Personal Perspective

The overexploitation of marine resources is a major global environmental problem, overlooked by much of the public. While some threatened living marine resources, such as whales, have garnered a great deal of public attention, depletion of once abundant fishery resources is virtually unknown outside of a rather small community of scientists, policymakers, public interest advocates, and those participating in fisheries both commercial and recreational. For most people, fish is available in the market or restaurants, albeit at rather high prices, so overfishing seems not to be all that widespread.

Surveys of public opinion find that most people in the United States believe pollution, not fishing, is the greatest threat to marine resources. Of course pollution is a major environmental problem, but usually a second-order effect. That is, fishing directly increases the mortality rate of fish or other exploited marine species, whereas pollution may reduce growth, reproduction, or survival in more subtle ways than simple removals. Other anthropogenic impacts such as coastal development, habitat loss, and climate change also impact living marine resources, again through second-order effects on productivity. But the existence of these other impacts does not negate the importance of fishing and overexploitation in depleting fish stocks and fundamentally altering marine ecosystems. Human activities are not independent of one another in their effects on changing ecosystems and in loss of natural productivity. Unfortunately, the impacts exacerbate one another. So, depletion due to fishing may occur more quickly because of habitat loss, and habitat loss in turn may be partially due to fishing activities.

For many populations of fish and other organisms, as abundance declines from presumably high levels before fishing occurred, populations compensate for the fishing pressure by increased production through growth and reproduction. This is the basis for much of the theory of fishing and for the expectation that some level of exploitation is sustainable for many biological populations. Overfishing occurs when the rate of mortality due to fishing is so high that the population cannot compensate with increased production so that abundance continues to decline. Fishing at these high rates becomes unsustainable. More complex effects occur when fishing changes age structure, habitat, or subpopulation structure.

Unfortunately, we have had ample opportunity to observe the effects of overfishing, both simple and complex. It has long been known that fisheries in general have a tendency toward overexploitation due to the overall economic and business pressures of the industry. Some spectacular stock collapses have occurred and, unfortunately, more are likely to occur in the near future. Pauly, Myers, Jackson, and others have documented widespread declines of fisheries resources worldwide. Though these analyses are controversial with respect to technical details, the overall pattern that arises from the fundamental character of the data on fishing and marine ecosystems is clear: very large declines in the abundances of many marine populations in virtually every ocean.

There is no better example of the results of overfishing than consideration of the Atlantic cod stocks. Cod have been a mainstay of fisheries for much of the developed world for more than five hundred years. Nonetheless, even with relatively simple hook and line gear, noticeable depletion of cod stocks occurred. From a variety of data sources, archaeological, historical, and ecological, it is clear that cod played an important role as predators within the Gulf of Maine ecosystem. The changing abundance of cod due to overfishing has had a major impact on trophic structure of the ecosystem through time.

Other commercially and ecologically important species such as lobster, urchins, and kelp respond to the changing predation pressure of cod. The long-term intensive overfishing of cod has not only reduced its abundance, but shifted the ecological balance in the Gulf of Maine with apparent trophic cascades affecting the whole ecosystem. Of course, the impacts of factors other than fishing, notably climate, also affect the ecological balance in this and many other areas. But the changing abundance of cod and the consequent effects on the ecosystem are most likely the result of the cumulative effects of fishing, climate, and other factors, as recently shown by Rose for northern cod stocks. A major issue in considering such cumulative impacts of overfishing and other changes in the ecosystem is whether overall system productivity has changed. In other words, if cod biomass was very much higher and trophic structure quite different before massive overfishing, has that productivity simply moved to other components of the ecosystem? Is productivity less today than in the past? In either case is the process reversible? If overfishing ends, what is the eventual "rebuilt" state of a system like the Gulf of Maine?

Moving beyond the biological effects of overfishing, resource depletion affects the social and cultural resilience of fishing communities. Overfishing is not a one-time single event for cod stocks in particular and fisheries in general. Rather, cod have been repeatedly overfished or perhaps continuously overfished for generations. During each period of overfishing, the communities dependent on the fishery adjusted to the changing resource condition by employing different gear, switching target species, and the migration of labor out of the community. Notably, the result of overfishing was rarely, if ever, the cessation of fishing and, therefore, a recovery period for the resources upon which the community depends. Throughout it all, governance of the fishery, hamstrung by political pressures, has lagged behind the actual impacts, often to such an extent that the overfishing has been nearly unaffected or even exacerbated by management. For these reasons, historical analyses are far more than just of academic interest. They provide a perspective that is lost by looking only at the recent history of a fishery or an ecosystem.

Until very recently, fisheries science and, to an even greater extent, management has paid little attention to the history of fisheries and the data that can be developed through historical research. From a scientific perspective, the emphasis has been on analysis and assessment of biological processes. The accuracy and precision of data sources and the development of more sophisticated analytical methods have dominated scientific advice for the near-term problems of management such as catch quota setting. Management plans intended to end overfishing and rebuild resources almost never used the lessons of history as a guide until quite recently. Frankly, managers have had enough of a struggle trying to reverse the downward spiral of fish stocks, let alone rebuild fully functional ecosystems. But looking only at recent patterns in a fishery exacerbates the problems of continually overfishing a resource. That is because recent patterns from a heavily exploited fishery hide the real potential of the species if exploitation is truly reduced. The historical analysis allows us to consider what might be possible if we change the way we exploit an ecosystem.

History has other lessons to teach us, too. It emphasizes the ignorance of repetitive overfishing. We can readily see that, in the Northwest Atlantic cod fisheries as with many others, we replaced overfishing by foreign fleets in U.S. and Canadian waters with overfishing by domestic fleets very quickly. Somehow, fishery management policy accepted the absurd premise that the only real problem was that "they" were overfishing rather than "we" and that the resource wouldn't crash for us even if the fishing mortality rate was the same as it was *before* the foreign vessels left. Historical analyses also clarify the magnitude of the changes that must take place to end the current overfishing. If current biomass of an enormously productive species like cod is 5 percent of the biomass 150 years ago, then making a minor reduction in fishing rates in the Northwest Atlantic or the North Sea is unlikely to result in real recovery for "decades or centuries," if at all. And history shows us that the social and economic losses due to repetitively overfishing a resource are far greater than the short-term losses due to reducing fishing pressure now. The political battle is usually over the immediate near-term impact of restrictions and ignores the fact that the lack of restriction results in long-term economic and societal costs that can be very large. Maine's island communities are a good example. When fishing was good, many islands supported healthy year-round populations, but now most are left to summer people.

While such losses are out of sight and out of mind for most of the public, marine resources and marine ecosystems are held in the public trust. Internationally, the phrase in the Law of the Sea is that these resources are the "common heritage of mankind." Allowing overfishing to occur gives up that heritage and violates the public trust. Even if most in the body politic are unaware or unconcerned about overexploitation, the need for wise stewardship is still there.

The political pressure with regard to fishery issues mostly comes from those immediately affected by restrictions, that is, the fishing community and businesses, recreational or commercial. That pressure is almost invariably opposed to restrictions, not from any intent to deplete resources, but rather to maintain maximum opportunity and flexibility for fishers. A commercial fisherman wants to maximize the chance to be successful in earning a good living from the fishery. A recreational fisherman wants to maximize the chance of enjoying a day's fishing. But collectively they can exert enormous pressure on a natural resource. Declining fish stocks are putting the interests of commercial and recreational fishers at odds because they have to compete for smaller portions of the total catch. Trophic interactions also occur among fisheries. For instance, commercial trawlers that intercept large schools of herring before they can come near shore have the secondary effect of driving popular recreational species like striped bass that prey on herring, thus having an adverse impact on the recreational fishery.

The power of fishing vessels is remarkable. A modern vessel with GPS, color sounding equipment, large engines, and extraordinarily strong synthetic materials used for fishing gear can locate and exploit any part of the ocean. Note that all of the features listed above are available for recreational as well as commercial vessels and have a consummate affect on fishing power even if the scale of catches is quite different. If all the users are trying to maximize opportunity, then what happens to that public trust? Elected officials respond to the political pressure of the loudest voices engaged in an issue. Then regulatory actions must be modified by that same pressure, at least in a democratic system. The result has been an excruciatingly slow response to the problem of overfishing and the loss of too many resources. Despite differences in political systems around the world, this pattern is quite widespread.

In recent years, public interest or environmental groups have entered the fray on the side of regulating fishing. The reaction from commercial and recreational fishing groups, and, sometimes, elected officials or governments, has been predictably negative, sometimes eliciting cries of horror at the attention from news media and increased litigation in U.S. courts from ostensibly well-funded environmentalists. Curiously, most lawsuits were and probably still are brought against the government by fishing groups for allegedly overzealous or inappropriate regulation. However, there was little outcry against litigation until a few successful suits by environmental groups resulted in *more* aggressive regulations. In spite of the hue and cry, recent litigation has raised some interesting issues. Some have successfully challenged the process for determining appropriate regulations under the National Environmental Policy Act. Difficult questions have been raised in some of these cases. To what extent have cumulative impacts on marine ecosystems been analyzed and considered in the decision on regulatory actions? Have the options considered covered a broad enough spectrum of choices? Clauses in primary fisheries statutes calling for broader ecosystem-level habitat protections and the reduction in bycatch have been a major focus of litigation. Bycatch occurs when a directed fishery catches nontarget fish. Often bycaught fish cannot be sold and are discarded at sea or at a processing plant. Thus, the ecosystem effects are considerable.

The legal arguments are beyond our expertise. However, the concept of cumulative impacts and ecosystem-level effects involve difficult but crucial ecological questions directly related to the global pattern of overfishing. The cumulative effect of human impacts on any particular ecosystem or the global commons as a whole is clearly

1. *large*, such as a 95 percent decline in some resources;
2. *profound*, causing change such as major alteration of food webs; and
3. *severe*, resulting in the potentially irreversible loss of ecosystem goods and services, that is, the properties of ecosystems that we extract as goods like fish or need as services like carbon sequestration.

On the one hand, it is laudable that the statutes in the United States and other nations, and recent international instruments such as the U.N. Fish Stocks Treaty, recognize the need to conserve ecosystems and not just target species. On the other, having the statutory or treaty authority is still a long way from the implementation of management systems that truly protect against overfishing and conserve ecosystem goods and services. Hence, the pace of change in fisheries management is slow, despite the statutes, lawsuits, and public attention.

Cod is the poster child for the effects of overfishing. Every Atlantic cod stock has been overfished and depleted. The pattern of overfishing has been repeated again and again as if learning by example were anathema. Each fishery has been declared unique and then followed a pattern remarkably similar to all of its predecessors. Even within a fishery the battle to reduce fishing pressure is fought over and over on the same political grounds and with halting progress at best. Each slight upturn in the stocks is taken as a signal to ease the restrictions on fishing for one more year, one more dollar, or one more euro. Managers now negotiate over thousands of tons of biomass when historically most stocks were hundreds of thousands or millions of tons.

Once of huge ecological, economic, and societal importance across the North Atlantic, only remnant populations of cod are left, a small fraction of the abundance that existed only a century ago. One measure of the importance of the species is that fishing still continues today on those remnant stocks of cod. Some of the remaining fisheries, such as in Iceland, may even be sustainable, but relative to historic productivity, at the current very low levels. In other cases such as on the Scotian Shelf and George's Bank, some slow rebuilding is occurring, but rebuilding targets are only to those same minimal levels seen in recent times. Off Newfoundland and Labrador, even under an albeit incompletely closed fishery, little rebuilding of the once great stock of northern cod is evident. And in the North Sea, the continuing fishery has driven the stocks down to low levels of biomass never before observed. It is almost as if we are addicted to fishing for cod even when the remaining stocks are on life support. Maybe the enormous productivity of the ocean that was once manifest in the cod stocks of the North Atlantic can come back to feed future generations, but not without efforts toward rebuilding the remaining resources that are as concerted as the efforts people have made to overfish these stocks again and again.

Shifting Baselines of Management

As Alan Longhurst has charitably observed, fisheries management is undergoing a period of introspection. Policymakers have employed the best available assessments and scored important victories against opponents of regulation, but historical stock declines of 80 percent or more across the board have not reflected well over the long term. A historical perspective can offer insights on this crisis in policy and governance.

Fisheries management is not new. Regulations likely sprang up as soon as overfishing caused hardships among fishers and markets. In the Middle Ages fresh- and then saltwater fisheries declined in northern Europe. Europeans responded by developing aquaculture, especially for carp, and by regulating fishing gear along the coast. Citing as rationale the destruction of fish stocks, the decline in average size, and the effect of poor quality and high-priced fish on food supply, royal edicts regulated gear and imposed closures along the French coasts by the fourteenth century to stabilize these valuable resources.

Fisheries regulation accompanied the Pilgrims to America. Three years after settlement, Plimouth Plantation gave itself the legal authority to regulate fisheries. Settlements begun in Piscataqua in 1623 were outside the Plimouth commonwealth and threatened its monopoly on marine resources, and the Pilgrims wanted the positive right via due process to protect their fish. Town laws also governed local fisheries, determining the physical location of fish flakes, dams, and weirs and licensing privileges to catch and preserve certain fish. Arguments pro and con are sometimes recorded, and will be familiar to anyone who has recently attended hearings on fisheries management plans. However, colonial fisheries management anticipated coastal zoning and nested layers of governance by four hundred years.

Regulations imposed by the Pilgrims and Puritans look very much like regulations today. They limited days fishing, established fishing seasons, protected spawning, issued licenses, imposed gear restrictions, and even instituted a landings moratorium—all before 1700. People rarely regulate abundant resources unless they anticipate the possibility of depletion. But the European settlers knew their history. Remembering exhausted Old World waters, and a shortage of cod near Boston in the 1650s, encouraged them to take a precautionary approach, protecting fish stocks they depended upon with the best available management practices. In the United States fisheries conservation contended with unfettered resource exploitation until after the Civil War, when laissez-faire policies favoring industrial fishing and factories won the day and undermined small producers. It is important to find out how effective regulations were in the past, and under what circumstances, but the lesson here is that a historical perspective often resulted in remarkably effective and adaptive risk management.

Shifting baselines promotes an unconscious preference for standards that compliment our frame of reference. For this reason, core scientific concepts deserve periodic reappraisal. What follow may be recent examples of shifting baselines at work in New England fisheries management.

Halliday and Pinhorn showed that the statistical areas used in New England fisheries management today were originally based on the distribution of cod and haddock, the most important commercial species in the 1930s, and adapted to other species after the fact. Retained because they are familiar and easy to use, these relatively large regions hide important variations in the geographic distribution of catch over time. For instance, in 1902, 162 million pounds of Atlantic herring were caught in coastal nets along the New England shore. In 2008, 165 million pounds were taken in statistical areas that extended past the continental shelf and covered half a million square kilometers. Clearly, catch distribution and density have changed in the past century, and this may have important implications for the health of Atlantic herring stocks.

Maximum sustainable yield (MSY) is an essential reference point in stock assessment models and was developed from a substantial body of work by a wide range of scientists. Many managers today view it as key to implementing a precautionary approach, although results have too often been problematic due to political pressure from economic interests. However, Carmel Finley recently pointed out that the adoption of MSY for fisheries management was less the product of deliberative peer review than a policy decision to expedite fisheries treaties in the wake of World War II. Ecological and economic deficiencies may have been overshadowed by the appeal of its "relative simplicity" and "lower information demand."

Subsequently, stock assessment models have been modified to improve estimates and forecasts. In 2007 Northeast Fisheries Science Center (NEFSC) scientists modified stock assessment models for Gulf of Maine cod. They went from a parametric approach, which yielded an MSY of 16,600 metric tons, to a nonparametric approach, which yielded an MSY of 10,014 metric tons. This recovery target is 40 percent smaller and easier to reach, although cod stocks are still moribund.

Also in 2007, NEFSC scientists modified the stock assessment models for monkfish. Old models employed a time series starting in 1963, even though demand was lackluster and data poor until Julia Child took up the cause in 1979. Based on the old system of reference points, monkfish were considered overfished in 2006. Under the new system developed in 2007, data before 1980 were discarded. Results indicated that monkfish were not overfished and overfishing was not occurring, although fish size had declined. According to Haring and Maguire, "NMFS [National Marine Fisheries Service] was considering more drastic measures to try to meet the rebuilding targets by 2009. The assessment results released in August 2007 suggest[ed] that this [would] not be necessary." The fishery's status quo was maintained. However, biomass in 2009 was less than the value projected in the 2007 assessment. The northern monkfish stock exhibited a strong retrospective pattern, that is, one informed by prior events. Although overfishing is still not projected to occur, biological indicators have been adjusted downward.

Maine lobsters are generally considered to be abundant and well managed by local cooperatives. Upper and lower size limits protect juveniles and breeding stock, spawning females are protected, and effective governance is nested—local decisions are augmented with input from state and federal agencies. If the cod fishery is the poster child for dysfunctional management, some propose that the lobster fishery be its role model. The stock has ostensibly increased to almost twice the management target and mortality levels remain acceptable, even though fishing effort is high despite hard economic times. Recent catch has been relatively consistent, and the cause of considerable pride.

Historical catch tells a different story. In 1887 Maine lobstermen averaged 200 lbs/trap, and in 2005 not quite 20 lbs/trap over the same grounds. Although trap limits were put in place in the 1990s, effort is as high or higher than ever before. The states have the lead in management, and they have not reduced fishing effort due to industry pressure. This is evident in the state of the southern New England lobster fishery, where the lobster population collapsed in 2003. Regulators are considering a five-year fishing moratorium to rebuild stocks.

Errors made because historical perspectives have *not* been taken into account are beginning to be identified and corrected. Yet finding the appropriate time frame for management actions is problematic. Everyone wants quick results, especially fishers chaffing under restrictions aimed at rebuilding fish stocks. Even if we agree that regulations work, we want them to work really quickly. But economic, biological, ecological, oceanographic, and climate cycles are simultaneously at play, and recovery depends on complex interactions at different temporal scales. Economic cycles run from year to year, but climate cycles may last for millennia.

MacCall found that outcomes of management models differed for short-lived and long-lived species, and for competitive species responding to fishing pressure under long periods of adverse climate. This means that policy measures should be developed for each species on a case-by-case basis in an ecosystem context and that responses to environmental changes should be carefully monitored. This is especially important in light of global warming. Lag time in implementing catch restrictions may actually benefit fisheries for short-lived species like herring, but for long-lived, cool-water predators like cod, recovery from overfishing under adverse climate conditions may take a century. This gives new meaning to staying the course and shows the scale of resolve necessary to truly rebuild fish stocks and ecosystems.

However, Longhurst questioned the very concept of sustainability, pointing out that there is no historical evidence that sustainable fisheries have ever been maintained for long periods of time.

Evidence of Progress

Put in practice, the shifting baselines paradigm challenges traditional perspectives on governance and scales of observation, refocusing management from single species to ecosystems, and acknowledging the role of humans as a key species. Several case studies in New England show positive results from this change in management perspective, often called spatial management.

Habitat Protection

The success of fisheries closures proved that protecting ecosystems protects species. Although controversial at the time, NMFS regulations closed more than 5,000 square nautical miles of prime fishing grounds beginning in 1994, in an attempt to restore depleted groundfish resources. Closure design focused on essential fish habitat (EFH), especially for cod, haddock, and yellowtail flounder. The five resulting fishery closures in the Gulf of Maine and on Georges Bank restricted commercial bottom-trawl fishing and scallop dredging, but they did not restrict all fishing. Generally, commercial fishing for herring and mackerel was permitted, with some restrictions to reduce groundfish bycatch, and all types of recreational fishing were allowed.

These closures helped to successfully rebuild Atlantic sea scallops and haddock on Georges Bank, although cod and yellowtail flounder are still recovering. Georges Bank haddock stock size increased tenfold from 1995 to 2005 partly due to a very strong year-class in 2003 and strong primary productivity. In 1999 high-resolution video surveys of about 5 percent of Georges Bank scallop grounds showed that density of the shellfish had increased to about half of the average harvest from 1977 to 1988. Ten years later, videos showed that Platts Bank, Fippennies Ledge, Jeffreys Ledge, and Cashes Ledge in the central Gulf of Maine also supported high densities of juvenile scallops. Although still recovering from overfishing, yellowtail flounder biomass appears to be at levels not seen since the late 1960s.

Marine protected areas (MPAs), such as the fisheries closures on Georges Bank, are one of the few strategies with proven success on an ecosystem level. This makes them important tools for ecosystem-based fisheries management, in which people's actions are managed to ensure healthy marine ecosystems and sustainable resources. Establishing an MPA involves an important social and ecological trade-off: the long-term loss of fishing grounds for the promise of biological conservation and healthy fish stocks in the future. Understanding human impacts on multiple spatial and temporal scales can provide the socioecological baselines required for successful ecosystem-based management. This is where historical perspective comes in.

Among the new tools being developed for multispecies spatial and temporal management is the Swept Area Seabed Impact (SASI) model, which assesses bottom conditions and the impacts of fishing gear on the bottom. George Brown Goode (1887) and Walter Rich (1929) assessed the effect of bottom conditions on different fisheries, based on sampling surveys and fishermen's information, and these data provide essential baseline habitat conditions to gauge the effect of widespread dragging and bottom trawling. Acknowledging the importance of such historical baselines, the Stellwagen Bank National Marine Sanctuary Management Plan of 2010 was informed by an assessment of historical marine ecology and socially constructed fishing tradition.

Whale and Seal Management

The vast history of whaling and sealing and the collapse of marine animal populations worldwide are unnecessary to review here. Whale and seal conservation are success stories in New England, despite the fact that shore whaling took place on Cape Cod as late as 1900. Fishing moratoria and protecting nursery grounds through the Marine Mammal Protection Act of 1972 and the Endangered Species Act of 1973 allowed populations to rebuild and revisit areas in the Gulf of Maine where they were once common. Humpback whales now regularly feed off the New England coast, and seal populations have become large enough to attract sharks. Today the greatest risks to whales in New England waters are from boat strikes and abandoned fishing gear.

Stellwagen Bank National Marine Sanctuary was established in 1992 partly as a protected area for whales and seals. Resident and migrant mammals include five species of seals (harp, gray, harbor, hooded, and ringed seals), ten species of whales (humpback, minke, fin, sperm, right, beluga, orca, sei, blue, and pilot), along with white-beaked, white-sided, common, bottlenose, and Risso's dolphins and harbor porpoise. Whale watching, popular with New Englanders and tourists alike, generates considerable economic benefit for small businesses.

River Herring Management

River herring, the collective name for alewives and blueback herring, have declined dramatically in the past fifty years. Commercial catch went from approximately 70 million pounds in 1957, to 13.7 million pounds in 1985, to under a million pounds in 2007. Some scientists fear a total collapse is under way. Important to both freshwater and oceanic ecosystems, these fish undertake extensive migrations from coastal oceans to spawn in fresh and brackish water, running through a gauntlet of man-made obstacles and dangers. Then they return to the sea where they spend the majority of their lives.

Historical landings data support arguments to protect river herring populations that have been decimated by dam building, overharvest, and, most recently, bycatch in other commercial fisheries. Methods such as dam removal, fisheries closures, spawning closures, and bycatch monitoring and regulation integrate very long-term changes in ecosystems, human behavior, and the life histories of river herring to an extent not seen before in fisheries management. For instance, removing the 160-year-old Edwards Dam on Maine's Kennebec River in 1999 was controversial at the time, but it helped reestablish a spawning population of two million alewives in just ten years, as well as improve other aspects of the environment. Its success has encouraged dam removals elsewhere.

Yet commercial river herring landings in state waters are a historic low since 1887, and NOAA declared alewives and bluebacks "Species of Concern" in 2006. Historical data supported Amendment 2 to the Shad and River Herring Management Plan (2009), which includes a default closure of directed commercial fisheries in state waters by 2012 unless sustainable harvest plans have been approved.

Because they often mix with schools of Atlantic herring and mackerel, significant river herring bycatch occurs in the herring and mackerel fisheries, among others. Roughly 64,000 pounds of river herring have been taken in a single tow by large vessels fishing for sea herring, although the magnitude of bycatch likely varies because of unique at-sea migration patterns of genetically distinct stocks. In some places it might be the most significant factor driving declines, whereas elsewhere it might be negligible. Still, significant genetic variation within these species has been lost since historical times, and bycatch is a continuing threat today.

National Standard 9 of the Magnuson-Stevens reauthorization mandated that Fisheries Management Councils minimize bycatch to the extent practicable. The New England and Mid-Atlantic Fishery Management Councils are addressing river herring bycatch through management plans for targeted species by developing Amendment 5 to the Atlantic Herring Fishery Management Plan (NEFMC) and Amendment 14 to the Squid, Mackerel and Butterfish Plan (MAFMC). Spatial and temporal management is key to protecting these vulnerable fish, particularly before spawning. The locations and timing of river herring hot spots in New England waters have been modeled using recent data and confirmed by historical information. Thus, historical perspectives strengthen arguments for implementing strong management plans to address river herring declines, and in the case of dam removal, progress to date reflects the promise of the past.

New Directions

Early U.S. Fish Commission reports show that fishermen and scientists were not always at odds. Fish commissioners had great respect for local ecological knowledge given willingly by fishermen who were eager to work with the commission. In that spirit, a new emphasis on local ecological knowledge brings fishermen, scientists, and managers together to restore fish, fisheries, and fishing communities. For example, ongoing development of a spawning closure for cod in Ipswich Bay, a historically important spawning ground, includes scientists and fishermen in the process. History can provide a working model and act as a bridge between stakeholders who are often at odds.

National Standard 1 of the Magnuson-Stevens Act of 2006 required ending overfishing by setting annual catch limits by 2011. At the same time, New England groundfish management introduced a catch share system. Commercial fishermen receive an allocation of the catch limit, the total allowable catch (TAC). They may consolidate their allocation with other fishermen in a fishing sector, or retain it separately and remain in a common pool.

Since vessels have a maximum range, catch shares may affect spatial management plans. Fisheries operating near closure boundaries receive the benefit of "spillover" productivity, but displaced fishing effort that concentrates along those boundaries may cause local depletion of fish stocks. Since fish were allocated, not fishing area, will fishermen become stewards of the closures that support productive populations, or will they exploit them for short-term gain?

Spatial management has other implications for governance. Steneck and Wilson argued that, since ecological and social processes operate simultaneously on different spatial scales, management at multiple levels is needed. Nested layers of management creates the checks and balances necessary to protect complex ecosystems, reinforce local knowledge systems and incentives for stewardship, and limit perverse incentives to overharvest.

Scales of governance are even more important now that coastal oceans are subject to conflicting usage. Ocean transportation carries much of what affluent consumers eat, wear, and use daily across sea-lanes and through fragile coastal waters. Underwater cables thread the ocean floor, and wind farms and tidal generators join offshore drilling platforms to produce energy. Commercial and recreational fisheries, and recreational, aesthetic, and conservation interests now compete with industries for prime ocean locations. Even waste disposal, with its particularly long, unsavory history, has become impossible to ignore any longer. Ocean zoning attempts to rationalize competing needs and interests while maximizing human well-being and ecosystem integrity. Although many fishermen are wary, marine scientists are coming to see its value. However, the Bluewater Horizon in the Gulf of Mexico and the Cape Wind Project off Cape Cod prove how extraordinarily difficult managing competing interests will be.

Loss of ecosystem services has been linked to loss of ecosystem resilience, economic viability, and human security. Tracking and assessing changes in space and time across multiple scales is a daunting task, but it has great value in modern ocean governance and will be increasingly necessary in light of the global and local changes taking place. As we confront the future, looking backward may be almost as important as looking forward.

Chapter 11

Lessons from Coral Reefs

ENRIC SALA AND JEREMY B. C. JACKSON

The collapse of human societies has commonly involved unsustainable overexploitation of resources and rapid population growth, followed by an environmental catastrophe, such as a prolonged drought, that destroyed the remaining resources. Collapse was sometimes averted by expansion into new territories to tap unexploited resources, which served as spatial subsidies that fueled further population growth. Finally, when expansion was no longer possible, collapse was even more sudden and severe (figure 11.1).

Collapse of marine ecosystems and fisheries has followed a strikingly similar pattern (figure 11.1). Fisheries managed to achieve "maximum sustainable yield" are especially vulnerable to environmental disturbance. Most are already at or below their lower limits of productivity. Increased fishing capacity supported by new technologies or economic subsidies may help to maintain or even increase catches in the short term, as in the case of cod. Then, when the fishery finally collapses, fishers move on to other species, which are generally smaller and grow faster. The economic collapse of one species ripples throughout the ecosystem as others become sequentially overexploited. No one accepts responsibility for these mistakes. Without historical memory, few even remember them. So the cycle of overfishing is repeated over and over until there are almost no fish left.

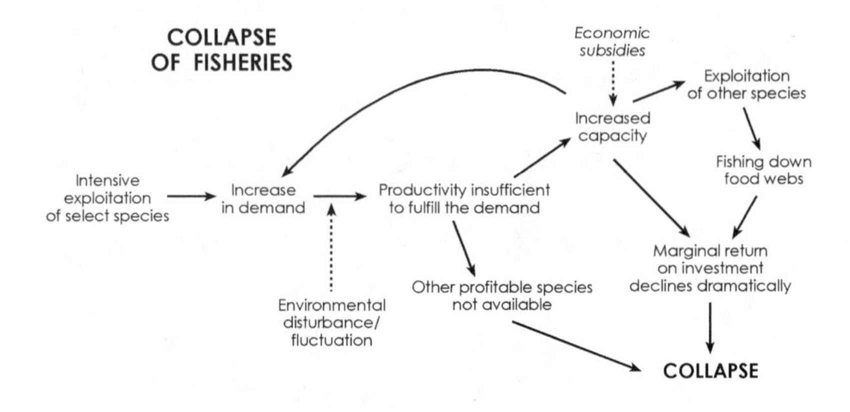

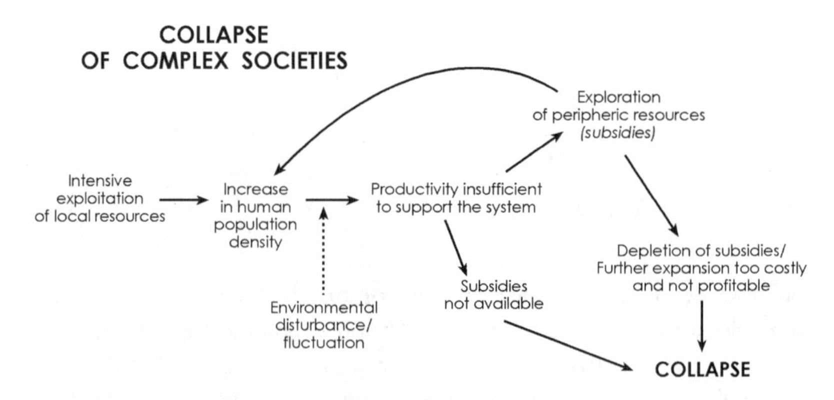

FIGURE 11.1. Historical sequence of events in which overexploitation of local resources and environmental disturbance may lead to the collapse of societies and ecosystems. Access to distant resources may delay the collapse of the system at the risk of decreasing the future possibility of recovery.

Atlantic bluefin tuna provide a fascinating example of this shifting baselines problem in fisheries. Every year these giant fish migrate throughout the Mediterranean to reproduce. Numbers that are unimaginable today were caught in giant *almadraba* traps, an ancient technique of setting nets in a maze to capture the tuna in a central pool. Aristotle witnessed waters boiling with a moving tide of giant tuna. In the Middle Ages, the same technology was still being employed, and in the mid-1600s a single trap on the southern coast of the Iberian Peninsula caught up to 100,000 tuna every year. There were hundreds of such traps throughout the Mediterranean, and catches began to decrease several centuries ago (figure 11.2). The

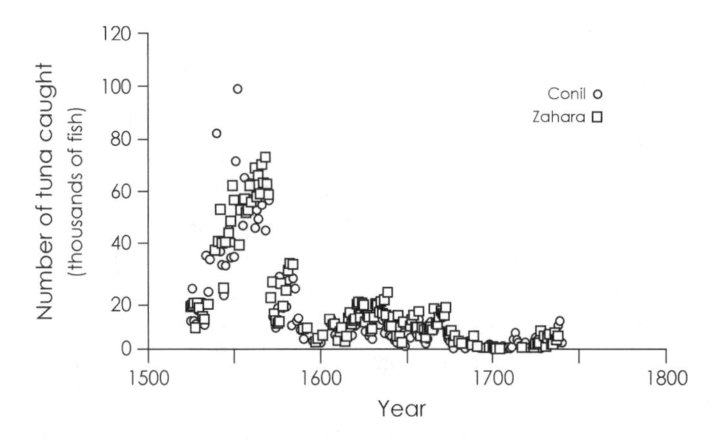

FIGURE 11.2. Catches of bluefin tuna (*Thunnus thynnus*) in two sites in southern Spain, from 1525 to 1740. Note that small periodic fluctuations that could be attributable to variation in environmental factors are negligible compared with the larger decline after 1550, presumably caused by fishing. (Unpublished data from the archives of the Medina-Sidonia family.)

unambiguous pattern of decline is emblematic of what must have occurred for overfished cod, although the data are more fragmentary and recent. It also contrasts strikingly with the long-term oscillations in sardines and anchovies driven primarily by oceanographic change. Today, only a handful of *almadrabas* remain because the Atlantic tuna population has been overfished to near extinction and is listed as "critically endangered" by the International Union for the Conservation of Nature (IUCN). Nevertheless, modern fisheries managers still highlight the importance of factors other than fishing, and the International Commission for the Conservation of Atlantic Tunas (ICCAT) still assumes that the few remaining tuna can be fished to achieve a maximum sustainable yield. Most egregiously in 2010, the Convention on International Trade in Endangered Species (CITES) caved in to intense lobbying and voted down the proposal that the species be listed under Appendix 1 of its commercial regulations, a measure that would have banned international trade in this endangered species.

The extreme degradation of ecosystems today seemed inconceivable a century ago when the ocean's riches were viewed as inexhaustible. Yet even then, the waters all around New York City were a dead zone. Human and animal sewage and industrial filth, and fishers with simple hook and line, harpoons, or traps along the shore, had inflicted massive impacts on marine ecosystems, now long forgotten.

Shifting Baselines on Coral Reefs

Just as for fisheries like cod, the lack of a baseline for pristine coral reefs plagues our understanding of their global degradation and limits our ability to do much about it. The Late Pleistocene fossil record tells us that coral communities exhibited remarkable stability in species composition and relative abundance of dominant species for several thousand to hundreds of thousands of years before their rapid demise in the twentieth century. However, modern ecological survey data are sparse until the 1970s when reefs were already mildly to severely degraded. Since the 1970s, live coral cover has declined by up to 50 percent in the Pacific and 90 percent in the Caribbean, with seaweeds commonly taking over from the corals.

Abundance of reef fishes has declined by 90–95 percent just in the last fifty years, and sea turtles by more than 99 percent. Comparisons of the very few relatively pristine reefs still remaining with the more exploited reefs yield great differences in coral and fish populations. However, sea turtles today are nowhere more than a small percentage of their abundance a century ago, and all are officially listed as endangered species by the IUCN. Monk seals are extinct in the Caribbean and on the verge of extinction in the Mediterranean and the Hawaiian Islands.

The known causes of coral reef decline include the direct and indirect effects of overfishing, pollution from the land, coastal development, and climate change, among myriad other factors. However, the interactions of all these factors are of such Byzantine complexity that it is difficult to untangle their comparative effects. Overfishing may greatly reduce some species at the expense of others, setting off chain reactions of events that alter the competitive balance between major groups of organisms. For example, loss of grazing fishes allows seaweeds to rapidly increase at the expense of corals. Seaweeds may kill corals directly by overgrowth or indirectly by promoting coral disease. The latter occurs because seaweeds leak vast amounts of organic matter into the surrounding seawater, as is obvious from the "smell of the ocean" along rocky coastlines in places like New England or Oregon. The organic matter provides food for countless bacteria, including coral pathogens, hence outbreaks of disease.

Runoff of sediments, toxins, excess nutrients, and sewage also kills corals directly or indirectly via increased microbial populations, loss of oxygen, and disease. Outbreaks of disease in turn cause catastrophic declines in formerly abundant sea urchins and corals, confounding the effects of overfishing and pollution. Finally, increases in atmospheric CO_2 are warming the oceans, which is a major factor in coral bleaching. Warming breaks down the symbiosis between corals and the microscopic algae within their tissues that provide most of the coral diet and are essential for coral growth. Sustained rises in temperature of only one to two degrees above normal highs can cause massive bleaching and mass mortality of corals. Increases in CO_2 are also making the oceans more acidic, which hinders coral growth and may dissolve their skeletons. Coral reefs as we know them may simply disappear.

No wonder there is so much disagreement and confusion about what to do. The problem is made even worse by lack of a historical record of when degradation began, which we are only now beginning to piece together. For example, old photographs reveal that the massive loss of formerly dominant staghorn and elkhorn corals was nearly complete in Barbados in the early twentieth century, and early qualitative surveys show comparable declines at sites scattered throughout the Caribbean before the 1980s. But elsewhere, dense populations of these corals persisted until the 1980s or 1990s when coral mortality due to overgrowth by seaweed, disease, and bleaching increased to epidemic proportions. The causes of these more recent losses are well documented, but have little bearing on the causes or magnitude of changes that may have occurred earlier elsewhere, as in the case of Barbados, and long before ecologists began to study them. Circumstantial evidence suggests that the problems in Barbados were due to deforestation of the island for sugarcane and the consequent runoff of sediments and human waste, as well as extreme overfishing to feed the burgeoning population.

The crucial points are that all the different causes of reef degradation are important and that degradation will continue and likely accelerate unless we reduce all the threats as soon as possible. If we could somehow wave a magic wand to halt climate change tomorrow, coral reefs would still likely disappear in a few decades unless we also halted overfishing and the runoff of pollution from the land.

Setting Goals for Conservation and Management

Historical ecology tells us what ecosystems were like before we overexploited and otherwise abused them and, therefore, implies what we could hope to achieve if we decided to restore them. Moreover, the trajectory of historical degradation provides a road map for restoration if we can reconstruct conditions as they were before. There is much talk and excitement about marine reserves as tools to restore biodiversity and healthy ecosystems, and perhaps to increase fisheries yields. But what does it mean to "restore" biodiversity, and how and when would we know whether we had succeeded or failed? Moreover, even if it were possible, do we as a society really want to return ecosystems to their pristine condition, and would we be willing to pay the enormous price?

In the case of the Florida Keys, would we be satisfied if reef fishes and corals became as abundant as they were in the 1970s? Or the 1950s? These would be wonderful achievements to be sure, but sharks, sea turtles, and manatee were already rare by the 1950s, sponges had suffered catastrophic declines due to disease, and the monk seal was already extinct. So do we need these creatures back as well, to the extent we can bring them back? Some people might not want superabundant sharks, but there is strong evidence that the extreme overfishing of sharks as well as other apex predators may destabilize reef food webs and community structure. Likewise, the extirpation of green turtles has been strongly linked with turtle grass wasting disease that wipes out entire meadows and their associated shrimp and fish stocks.

Historical trajectories of reef degradation can offer partial answers to these questions. For example, how do we know when a coral reef is in trouble and when drastic action is required to stave off disaster? That might seem obvious for the Florida Keys today, where fish, conch, and lobsters are severely depleted, most reef corals are dead from disease, bleaching, and overgrowth by seaweeds, and water quality has declined severely. But the collapse of Jamaican coral reefs was far from obvious before the one-two punch of Hurricane Allen and the mass mortality of the sea urchin *Diadema antillarum*—although with hindsight the extreme overfishing of reef fishes to the point of ecological extinction should have sounded the alarm.

Hurricane Allen struck Jamaica in 1980—the first bad storm in nearly a century. Immediately after the storm, scientists at the Discovery Bay Laboratory were back in the water to document the damage. This was the first time that the underwater effects of a hurricane were documented in such detail. Fifty-foot waves had wiped the reefs clean down to 30 feet or more, piling dead and stinking coral rubble into new islands on the reef crest and causing mayhem far below.

Hurricanes are nothing new in the Caribbean, and scientists confidently predicted how the reefs would recover based on the biology and vital statistics of the dominant coral species. But they got it all wrong. Two years later, a mysterious plague wiped out 95 percent of the ubiquitous sea urchin *Diadema antillarum*, which turned out to have been the last important grazing animal on Caribbean reefs. Within a few weeks the reefs were covered by a thin coating of filamentous algae and a few years later were carpeted in thick growths of seaweeds that smothered most of the remaining corals. Live coral cover plummeted from about 60 percent to less than 5 percent. Coral bleaching and disease have further inhibited coral recovery. Today, the waters are murky, and once beautiful reefs are mountains of seaweeds and slime.

On healthy coral reefs, grazing fish like parrotfish and surgeonfish are the most important grazers upon seaweeds along with invertebrates like the urchins. In the sparsely inhabited Northern Line Islands in the middle of the Pacific, grazing fish are still abundant and the corals there are some of the healthiest in the world, despite episodes of coral bleaching. On Jamaica, however, overfishing had upset this balance by the early 1900s. Thus, the fate of Jamaican reefs was sealed long before Hurricane Allen, pollution, and climate change.

We can use such historical trajectories to rank reefs along a gradient of degradation, identify reefs at risk, and identify conservation priorities to ward off imminent collapse. This is, in effect, what the government of Australia did in 2003 when it rezoned the Great Barrier Reef and closed one-third of the reef and all the different reef habitats to any form of exploitation, and more recently the designation of monument status for the entire northwest Hawaiian Islands by the U.S. government.

How will Australia and the United States know whether their bold actions have been successful, and what criteria should we use to evaluate success or failure? If we want to determine whether a marine reserve is fulfilling its conservation goals, we can monitor sites inside and outside the reserve over time. We can then incorporate the data into a historical database and do a new analysis to observe the ecological trajectory of the reserve sites over time. If the goal of the reserve is to recover ecosystem health, then the trajectory of degradation should be approximately reversed (figure 11.3A). We say "approximately" because environments and ecosystems are always changing for a variety of natural and human reasons besides the ones we pay attention to, so the world is inevitably in a very different condition than when degradation began. Thus, intervention may halt decline but result in an entirely different trajectory to a new alternative community state (figure 11.3B). As Thomas Wolfe famously said, "You can't go home again." Nevertheless, and regardless of this fundamental uncertainty, if the trajectory of degradation within a reserve is similar to unprotected sites or does not change (figure 11.3C), the reserve would not be fulfilling its stated purpose and additional actions would be required.

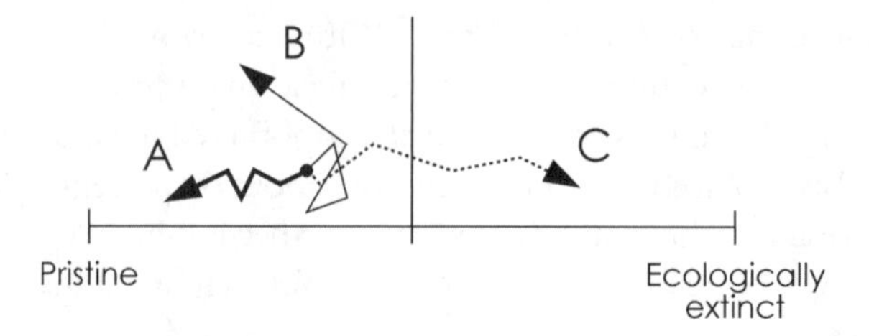

FIGURE 11.3. Hypothetical ecological trajectories of a protected coral reef. The phase space of the diagram represents the range of possible community composition from pristine to ecologically extinct in ordination space (sensu Pandolfi et al. [2003]; Pandolfi et al. [2005]). Intervention is begun when the reef community has degraded approximately one-third of the way en route to ecological extinction. Three possible trajectories are illustrated: (A) protection is successful in reversing the trajectory of degradation; (B) protection halts degradation but results in some new alternate community composition; and (C) protection fails to prevent further degradation and effectively makes no difference compared to unprotected sites. (Knowlton [2004]).

Challenges for the Future

Five fundamental questions need to be resolved if historical ecology can live up to its potential. Questions 1 and 2 are purely historical. Questions 3 and 4 combine historical analysis with ecological observations. Question 5 is the most important, and will require the combination of historical ecological perspective with appropriate large-scale and bold management initiatives and experiments.

1. *How variable were pristine marine communities and ecosystems over different spatial and temporal scales?* This question addresses the fundamental properties of baseline conditions that must have varied considerably among environments and ecosystems. Coral reef communities were remarkably stable for millennia until their geologically instantaneous collapse. In contrast, sardine and anchovy upwelling ecosystems exhibited enormous cyclical, and therefore predictable, fluctuations over comparable time periods. We need more examples to calibrate our expectations with rigorous sampling to distinguish signal from noise. The rich core records of planktonic foraminifera and coccolithophores from the deep sea, as well as cores from estuaries, seem particularly promising for analysis of pelagic baseline communities. There are unlimited replicate samples and material, much of it al-

ready processed for paleoceanographic studies, which are ripe for this sort of paleoecological analysis.

2. *What were the chronologies and rates of change along trajectories of ecological degradation?* There are two crucial components to this question. First, was degradation gradual or in steps, and did rates of change remain constant or increase toward the present? We want to know whether trajectories are nonlinear at all scales, with a series of little phase shifts along the way, or whether everything happened all at once when some critical threshold was reached. We can only guess the answer for most ecosystems because we assumed stability until everything appeared to collapse at the end, like California sardines in the 1950s, Jamaican coral reefs after 1983, or the debacle of cod. Second, were changes synchronous globally or within large water masses, or were they geographically out of phase? If changes were synchronous, they may have an oceanographic as well as anthropogenic explanation. However, if they were out of phase in ways closely correlated with human activity and not with climate, then human impacts must have been primarily responsible. These questions could be addressed using time series from sediment cores and uplifted outcrops of coral reefs.

3. *What were the proximate causes or mechanisms of collapse?* In terms of the "straw that broke the camel's back," we need to know whether all straws are, or were, equally important, or whether some changes or events were more important than others in the loss of resilience leading up to catastrophic shifts. We suspect that the most frequent proximate causes are stochastically occurring extreme natural disturbances such as especially strong hurricanes or oceanographic regime shifts that push anthropogenically stressed ecosystems beyond the point of no return. We could get much of the answer to this question by careful analysis of the environmental conditions leading up to and concurrent with well-studied examples of ecological collapse. Other important measures could include retrospective analyses of possible decrease in fitness of organisms before collapse occurred. For example, did rates of coral growth or regeneration of injuries decline beforehand? We could definitively answer these questions for reef corals by detailed analysis of cores already collected for paleoceanographic studies of climate change.

4. *Can we identify simple and reliable proxies of imminent ecosystem collapse?* The simplest and most reliable proxies for effects of exploitation include the body size of exploited species and the distribution of biomass at

different trophic levels of the food web. The absence of large fish is a sure warning of disaster, as is the shift in proportional biomass from apex predators to herbivores or from vertebrates to invertebrates. We urgently need to compile values of these simple indices along well-documented gradients of ecological degradation so that they can be quickly and easily applied to other previously unstudied ecosystems. Other likely proxies include retrospective measures of growth rates and other measures of fitness as described for corals above.

5. *Will degraded ecosystems tend to retrace their earlier trajectories of degradation after human impacts are reduced or strike out in different directions?* The problem of the path of "recovery," already alluded to, is an inevitable problem of multiple ecosystem states (figure 11.3), exacerbated by the possibility of entirely new sorts of assemblages due to changes in the available species pool or environmental conditions unrelated to the original reasons for collapse. Stated another way, since all marine ecosystems are significantly degraded from their pristine state, could we somehow engineer their "recovery" to some new and "improved" ecological state? The only way to know is to do ecosystem experiments on large enough spatial and temporal scales to accommodate the life history characteristics and behavior of previously important species, and, with far greater difficulty, attempt to reverse degradation once it has occurred.

Coda

So far the evidence for recovery from the few bold experiments and management actions is mixed. Indeed, bold management is almost inevitably a bold experiment, whereas weak management or inaction is almost always a bad experiment. Large-scale closures of fishing grounds such as those on Georges Bank have resulted in the partial restoration of fish stocks, especially when managers were able to act quickly and on appropriately very large scales to ward off total collapse. Likewise, reductions in excess runoff of nutrients and eutrophication have resulted in improvements in water quality and partial recovery of benthic ecosystems and fisheries in Tampa Bay and the Black Sea. In contrast, substantial reductions of nutrients in the Baltic Sea have so far failed to achieve the desired results due to the storage of vast quantities buried in sediments that recharge excess nutrients to the ecosystem.

Such disappointments are hardly surprising given the centuries of human insults and the timidity of our response. Moreover, the numbers of serious, large-scale attempts to address extensive ecological degradation are very few and have only just begun. We are still applying a piecemeal approach, focusing on runoff of nutrients and toxins while ignoring overfishing, or worrying about climate change while ignoring everything else. More integrated and holistic programs offer the promise of immediate short-term gains in some directions that can help sustain the commitment to stay the course in others. For coral reefs, protection of large areas from fishing and runoff of nutrients may result in rapid recovery of fish stocks but little, if any, short-term recovery of corals, which take so much longer to grow and reproduce. But none of this will happen if we don't start now while we still have a chance.

Acknowledgments

We thank E. Ballesteros, R. Bradbury, P. Dayton, N. Knowlton, L. McClenachan, M. Hardt, A. Rosenberg, and D. Wesson for discussions about history and human impacts on ecosystems. The tuna data were brought to our attention by F. Riera. To all of them we are very grateful.

Epilogue: Shifting Baselines for the Future

JEREMY B. C. JACKSON AND KAREN E. ALEXANDER

No one needed to "manage" fish when people were few and fishing pressures were low. Indeed, as Andy Rosenberg is fond of pointing out, the concept of managing fish is fundamentally absurd. Fish can't read the regulations and they don't go to management meetings, people do. So we must manage people in order to restore the fish, fisheries, and ecosystems that we value.

Yet managing people is difficult. People have legal rights, economic interests, and differing notions of well-being that can impede or facilitate management practices. Denial of responsibility permeates fisheries folklore and even scientific literature about the decline of sea turtles, monk seals, whales, seabirds, and many food fish, which until recently has been attributed to everything but overfishing. But the concept of shifting baselines helps to neutralize denial by spotlighting evidence to the contrary.

We end by highlighting what gives us hope—substantial changes in human behavior can make a significant difference.

Several marine species have been brought back from the brink. Since the International Whaling Commission declared a moratorium on commercial whaling in 1986, many whale species have rebounded dramatically, to the point that several nations now clamor to reopen the fishery despite substantial loopholes for scientific and other forms of whaling by some signatories. The extent of recovery is currently debated, but everyone agrees that increases since protection have been great. Establishing large closures on Georges Bank in the 1990s resulted in such a substantial rebound of scallops and haddock that fishing boats now crowd the closure boundaries and there is pressure to reopen the grounds to fishing. Of course no one would care if the closures had not worked.

On the newly rezoned Australian Great Barrier Reef, coral trout populations are rebounding, with the possible ecological benefit of their increased predation on crown-of-thorns starfish, which have devastated reef corals in the past. Likewise, the Northern Line Islands escaped heavy fishing largely through benign neglect and today support close to the highest reef fish biomass anywhere. They also exhibit among the highest cover of reef-building corals and coralline algae in the central Pacific despite increasingly high temperatures and coral bleaching. Apparently, marine protected areas confer greater resistance to the effects of global climate change. Although absolute protection has proven more difficult for developing nations, alternative strategies based on long-standing local traditions for management have proven effective in Melanesia and elsewhere.

Recognizing shifting baselines is the first step toward creating new ways of thinking that reintegrate the past, present, and future. Not to dwell on our past failures or to imagine returning to some idyllic pristine state, but to better envision ways of living that can heal the wounds of the natural world while improving people's lives: that is the lesson of this book.

NOTES

Introduction

Page 1

the end of the last ice age: Maca-Meyer, N., A.M. González, J.M. Larruga, C. Flores, and V.M. Cabrera (2001), Major genomic mitochondrial lineages delineate early human expansions. BMC Genetics 2:13 (doi:10.1186/1471-2156-2-13); Goebel, T., M.R. Waters, and D.H. O'Rourke (2008), The Late Pleistocene dispersal of modern humans in the Americas. *Science* 319: 1497–1502.

larger than 100 pounds extinct: Burney, D.A., and T.F. Flannery (2005), Fifty millennia of catastrophic extinctions after human contact. *Trends in Ecology and Evolution* 20(7): 395–401.

less than five million people: This and subsequent global population estimates in this chapter come from the U.S. Census Bureau, Population Division (2010), Historical Estimates of World Population, in *International Programs*, assessed online at: www.census.gov/ipc/www/worldhis.html.

heaps of bones and shells along the shoreline: Henshilwood, C., F. d'Errico, M. Vanhaeren, K. van Neikerk, and Z. Jacobs (2004), Middle Stone Age shell beads from South Africa. *Science* 304: 404; Vanhaeren, M., F. d'Errico, C. Stringer, S.L. James, J.A. Todd, and H.K. Mienis (2006), Middle Paleolithic shell beads in Israel and Algeria. *Science* 312: 1785–1788; Rick, T.C., and J.M. Erlandson, editors (2008), *Human Impacts on Ancient Marine Ecosystems: A Global Perspective*. Berkeley and Los Angeles: University of California Press.

firmly established a few thousand years later: Diamond, J. (1997), *Guns, Germs and Steel: The Fates of Human Societies*. New York and London: W.W. Norton.

Sargon the Great, around 700 BCE: plate 229 in Strommerger, E., M. Hirmer, photographer, and C. Haglund, translator (1964), *5000 Years of the Art of Mesopotamia*. New York: Abrams.

an ever-growing demand for fish and fish products: Bekker-Nielsen, T. (2004), *Ancient Fish and Fish Processing in the Black Sea Region*. Copenhagen: Aarhus University Press.

Page 2

or they targeted fish that were harder to catch or farther away: McClenachan, L., M. Hardt, J. Jackson, and R. Cooke (2010), Mounting evidence for historical overfishing and long-term degradation of Caribbean marine ecosystems: Comment on Julio Baisre "Setting a baseline for Caribbean fisheries." *Journal of Island and Coastal Archaeology* 5(1): 165–169; Kennett, D.J., B. Voorhies, T.A. Wake, and N. Martinez. Long-term effects of human predation on marine ecosystems in Guerrero, Mexico, p. 103–124 in Rick and Erlandson, editors (2008); and Bourque, B.J., B.J. Johnson, and R.S. Steneck. Possible prehistoric fishing effects on coastal marine food webs in the Gulf of Maine, p. 165–185 in Rick and Erlandson, editors (2008).

Atlantic gray whale, hunted to extinction by the eighteenth century: Bryant, P.J. (1995), Dating remains of grey whales from the eastern North Atlantic. *Journal of Mammalogy* 76(3): 857–861.

said to be inexhaustible: Huxley, T.H. (1884), Inaugural address of the fishery conferences. *Fisheries Exhibition Literature* 4: 1–19; Bolster, W.J. (2006), Opportunities in marine environmental history. *Environmental History* 11: 567-597.

even the mighty cod showed evidence of steep decline: Rosenberg, A.A, W.J. Bolster, K.E. Alexander, W.B. Leavenworth, A.B. Cooper, and M.G. McKenzie (2005), The history of ocean resources: Modeling cod biomass using historical records. *Frontiers in Ecology and the Environment* 3(2): 84–90; p. 114–160 in Roberts, C. (2007), *The Unnatural History of the Sea*. Washington D.C.: Island Press.

oysters, shad, and alewives from New England to the Chesapeake Bay: Kirby, M.X. (2004), Fishing down the coast: Historical expansion and collapse of oyster fisheries along continental margins. *Proceedings of the National Academy of Sciences of the United States of America* 101(35): 13096–13099; McPhee, J. (2002), *The Founding Fish*. New York: Farrar, Straus and Giroux; Limburg, K.E., and J.R. Waldman (2009), Dramatic declines in North Atlantic diadromous fishes. *Bioscience* 59(11): 955–965.

sea turtles and monk seals from the Caribbean: McClenachan, L., J.B.C. Jackson, and M.J.H. Newman (2006), Conservation implications of historic sea turtle nesting beach loss. *Frontiers in Ecology and the Environment* 4: 290–296; McClenachan, L., and A.B. Cooper (2008), Extinction rate, historical population structure and ecological role of the Caribbean monk seal. *Proceedings of the Royal Society B* 275: 1351–1358.

Chemical pollution and invasions of nonnative species also increased: Latimer, J.S., W.S. Boothman, C.E. Pesch, G.L. Chmura, V. Pospelova, and S. Jayaraman (2003), Environmental stress and recovery: The geochemical record of human disturbance in New Bedford Harbor and Apponagansett Bay, Massachusetts (USA). *Science of the Total Environment* 313(1–3): 153–176; Carlton, J.T. (2003), Community assembly and historical biogeography in the North Atlantic Ocean: The potential role of human-mediated dispersal vectors. *Hydrobiologia* 503: 1–8.

entire estuaries and coastal ecosystems were devastated by 1900: Lotze, H.K., H.S. Lenihan, B.J. Bourque, R.H. Bradbury, R.G. Cooke, M.C. Kay, S.M. Kidwell, M.X. Kirby, C.H. Peterson, and J.B.C. Jackson (2006), Depletion, degradation, and recovery potential of estuaries and coastal seas. *Science* 312: 1806–1809; Petts, G.E., H. Möller, and A.L. Roux, editors (1989), *Historical Change of Large Alluvial Rivers: Western Europe*. Chichester and New York: Wiley.

local or regional problems were rapidly becoming global: Jackson, J.B.C., M.X. Kirby, W.H. Berger, K.A. Bjorndal, L.W. Botsford, B.J. Bourque, R.H. Bradbury, R. Cooke, J. Erlandson, J.A. Estes, T.P. Hughes, S. Kidwell, C.B. Lange, H.S. Lenihan, J.M. Pandolfi, C.H. Peterson, R.S. Steneck, M.J. Tegner, and R.R. Warner (2001), Historical overfishing and the recent collapse of coastal ecosystems. *Science* 293: 629–637; Myers, R.A., and B. Worm (2003), Rapid worldwide depletion of predatory fish communities. *Nature* 423: 280–283; Pandolfi, J.M., R.H. Bradbury, E. Sala, T.P. Hughes, K.A. Bjorndal, R.G. Cooke, D. McArdle, L. McClenachan, M.J.H. Newman, G. Paredes, R.R. Warner, and J.B.C. Jackson (2003), Global trajectories of the longterm decline of coral reef ecosystems. *Science* 301: 955–958; Lotze et al. (2006).

cheap energy from a seemingly endless supply of fossil fuels: Meadows, D.H., J. Randers, and D.L. Meadows (2004), *Limits to Growth: The 30-year Update*. White River Junction, VT: Chelsea Green Publishing Company.

the world's major industrial fisheries have crashed or are over- or fully exploited: Myers and Worm (2003); Worm, B., E.B. Barbier, N. Beaumont, J.E. Duffy, C. Folke, B.S. Halpern, J.B.C. Jackson, H.K. Lotze, F. Micheli, S.R. Palumbi, E. Sala, K.A. Selkoe, J.J. Stachowicz, and R. Watson (2006), Impacts of biodiversity loss on ocean ecosystem services. *Science* 314: 787–790; Worm, B., R. Hilborn, J.K. Baum, T.A. Branch, J.S. Collie, C. Costello, M.J. Fogarty, E.A. Fulton, J.A. Hutchings, S. Jennings, O.P. Jensen, H.K. Lotze, P.M. Mace, T.R. McClenahan, C. Minto, S.P. Palumbi, A.M. Parma, D. Ricard, A.A. Rosenberg, R. Watson, and D. Zeller (2009), Rebuilding global fisheries. *Science* 325: 578–585; p. 30 in Food and Agriculture Organization (FAO) of the United Nations (2009), *The State of World Fisheries and Agriculture*, 2008. FAO, Rome, online at: http://www.fao.org/docrep/011/i0250e/i0250e00.HTM

fewer and smaller fish: McClenachan, L. (2009), Documenting loss of large trophy fish from the Florida Keys with historical photographs. *Conservation Biology* 23(3): 636–643.

are increasingly full of mercury, dioxins, and PCBs: Stokstad, E. (2004), Uncertain science underlies new mercury standards. *Science* 303: 34; Mozaffarian, D., and E.B. Rimm (2006), Fish intake, contaminants, and human health: Evaluating the risks and benefits. *Journal of the American Medical Association* 296(15): 1885–1899; Burros, M. (2007), Industry money fans debate on fish. *New York Times*, online at: http://www.nytimes.com/2007/10/17/dining/17fish.html?_r=1&sq=Industry%20Money%20Fans%20Debate%20on%20Fish&st=cse&adxnnl=1&scp=1&adxnnlx=1289066470-4/Osh6MqFxJ4+6ME9Vbryg

plastics are trapped in ocean gyres: Derraik, J.G.B. (2002), The pollution of the marine environment by plastic debris: A review. *Marine Pollution Bulletin* 44(9): 842–852; Gramling, C. (2009), Voyage to the plastic vortex. *Earth* 54(11): 22–23.

a few dozen in the 1950s to more than four hundred today: Diaz, R.J., and R. Rosenberg (2008), Spreading dead zones and consequences for marine ecosystems. *Science* 321: 926–929.

bleaching and disease fueled by rising temperatures: Hughes, T.P., A.H. Baird, D.R. Bellwood, M. Card, S.R. Connolly, C. Folke, R. Grosberg, O. Hoegh-Guldberg, J.B.C. Jackson, J. Kleypas, J.M. Lough, P. Marshall, M. Nyström, S.R. Palumbi, J.M. Pandolfi, B. Rosen, and J. Roughgarden (2003), Climate change, human impacts, and the resilience of coral reefs. *Science* 301: 929–933; Donner, S.D., T.R. Knutson, and M. Oppenheimer (2007), Model-based assessment of the role of human-induced climate change in the 2005 Caribbean coral bleaching event. *Proceedings of the National Academy of Sciences of the United States of America* 104(13): 5483–5488.

threatens virtually all sea life with calcareous skeletons, including corals, shellfish, and plankton: Orr, J.C., V.J. Fabry, O. Aumont, L. Bopp, S.C. Doney, R.A. Feely, A. Gnanadesikan, N. Gruber, A. Ishida, F. Joos, R.M. Key, K. Lindsey, E. Maier-Reimer, R. Matear, P. Monfray, A. Mouchet, R.G. Najjar, G. Plattner, K.B. Rodgers, C.L. Sabine, J.L. Sarmeinto, R. Schlitzer, R.D. Slater, I.J. Totterdell, M.-F. Weirig, Y. Yamanaka, and A. Yool (2005), Anthropogenic ocean acidification over the twenty-first century and its impacts on calcifying organisms. *Nature* 437: 681–686; Kerr, R.A. (2010), Ocean acidification unprecedented, unsettling. *Science* 328: 1500–1501.

the open ocean pelagic realm is "threatened": Jackson, J.B.C. (2008), Ecological extinction and evolution in the brave new ocean. *Proceedings of the National Academy of Sciences of the United States of America* 105 (supplement 1): 11458–11465.

Page 3

the so-called global ecological footprint-from less than 50 percent in 1950 to 150 percent today: Wackernagel, M., D. Moran, and S. Goldfinger (2004), *Ecological Footprint Accounting: Comparing Resource Availability with an Economy's Resource Demand*. Oakland, CA: Global Footprint Network, online at: http://www.envirosecurity.org/conference/working/EFAccounting.pdf; Global Footprint Network (2003–2009), online at: http://www.footprintnetwork.org/en/index.php/GFN/page/2010_livingplanet_report/

because we ignore historical change and accept the present as natural: Pauly, D. (1995), Anecdotes and the shifting baseline syndrome of fisheries. *Trends in Ecology and Evolution* 10: 430.

estimate past changes and understanding those changes in a social and historical as well as scientific context: Jackson et al. (2001); Roberts (2007); Bolster, W.J. (2008), Putting the ocean in Atlantic history: Maritime communities and marine ecology in the Northwest Atlantic, 1500–1800. *American Historical Review* 113: 19–47.

lessons that must now be incorporated into fisheries management: Longhurst, A. (2010), *Mismanagement of Marine Fisheries*. Cambridge and New York: Cambridge University Press.

Page 4

"it is not possible to maximize simultaneously generality, realism, and precision": p. 7 in Levins, R. (1968), *Evolution in Changing Environments*. Princeton, NJ: Princeton University Press.

for degraded coral reefs, and for estuaries and coastal seas around the world: Jackson et al. (2001); Pandolfi et al. (2003); Lotze et al. (2006).

historical data is rarely precise enough to estimate past populations (although evidence mounts to the contrary): Rosenberg et al. (2005); McClenachan et al. (2006); Poulsen, R.T., A.B. Cooper, P. Holm, and B.R. MacKenzie (2007), An abundance estimate of ling (*Molva molva*) and cod (*Gadus morhua*) in the Skagerrak and the northeastern North Sea, 1872. *Fisheries Research* 87: 196–207; McClenachan and Cooper (2008); Alexander, K.E., W.B. Leavenworth, J. Cournane, A.B. Cooper, S. Claesson, S. Brennan, G. Smith, L. Rains, K. Magness, R. Dunn, T.K. Law, R. Gee, W.J. Bolster, and A.A. Rosenberg (2009), Gulf of Maine cod in 1861: Historical analysis of fishery logbooks, with ecosystem implications. *Fish and Fisheries* 10: 428–449.

Page 5

Oviedo cataloged not just these big animals, but also fish, sponges, lobsters, conchs, and sea cucumbers: Oviedo, F. de. (1535), *General and Natural History of the Indies*. Seville, Spain.

heretofore, scientists assumed there were fewer than twenty nesting sites: McClenachan et al. (2006).

restoring both the turtle populations *and* the seagrass ecosystems: McClenachan et al. (2006).

the recent global decline of large predatory fish: Myers and Worm (2003).

the collapse of the Newfoundland cod fishery after five hundred years of commercial fishing: Vickers, chapter 7, this volume.

Chapter 1

Page 15

the weight of big, carnivorous fishes was only 3 percent of the entire fish community around the main Hawaiian Islands, but was 54 percent in the remote Northwest Hawaiian Islands, and even more around Palmyra Atoll: Friedlander, A.M., and E.E. DeMartini (2002), Contrasts in density, size, and biomass of reef fishes between the northwestern and the main Hawaiian Islands: The effects of fishing down apex predators. *Marine Ecology Progress Series*230: 253–264.

here, top predators comprised 85 percent of reef fish biomass: Sandin, S.A., J.E. Smith, E.E. DeMartini, E.A. Dinsdale, S.D. Donner, A.M. Friedlander, T. Konotchick, M. Malay, J.E. Maragos, D. Obura, O. Pantos, G. Paulay, M. Richie, F. Rohwer, R.E. Schroeder, S. Walsh, J.B.C. Jackson, N. Knowlton, and E. Sala (2008), Baselines and degradation of coral reefs in the northern Line Islands. *PloS One* 3(2): e1548 (doi:10.1371/journal.pone.0001548).

Eye of the Albatross: Safina, C. (2002), Eye of the Albatross. New York: Henry Holt.

"in those twenty leagues, the sea was thick with turtles so numerous it seemed the ships would run aground on them and were as if bathing in them": Jackson, J.B.C. (1997), Reefs since Columbus. *Coral Reefs* 16(supplement): S23–S32.

"It is affirmed that vessels which have lost their latitude in hazy weather have steered entirely by the noise which these creatures make in swimming": Jackson (1997).

Page 17

There is excellent information that is still low fruit, waiting to be analyzed: See, for instance, the variety of information employed in Jackson, J.B.C., M.X. Kirby, W.H. Berger, K.A. Bjorndal, L.W. Botsford, B.J. Bourque, R.H. Bradbury, R. Cooke, J. Erlandson, J.A. Estes, T.P. Hughes, S. Kidwell, C.B. Lange, H.S. Lenihan, J.M. Pandolfi, C.H. Peterson, R.S. Steneck, M.J. Tegner, and R.R. Warner (2001), Historical overfishing and the recent collapse of coastal ecosystems. *Science* 293: 629–637.

Ram Myers and Boris Worm showed this in their analyses of existing, accessible data sets that had previously been ignored: Myers, R.A., and B. Worm (2003), Rapid worldwide depletion of predatory fish communities. *Nature* 423: 280–283.

showed a clear relationship between abundance of spawning fish and spawning success: Safina, C., and D.H. Klinger (2008), Collapse of bluefin tuna in the western Atlantic. *Conservation Biology* 22(2): 243–247.

Page 18

the shifting baselines problem: Pauly, D. (1995), Anecdotes and the shifting baseline syndrome of fisheries. *Trends in Ecology and Evolution* 10: 430.

we have very informative reconstructions from Indian middens: Jackson et al. (2001); Steneck, R.S., and J.T. Carlton (2001), Human alterations of marine communities: Students beware! p. 445–468 in Bertness, M.D., S.D. Gaines, and M.E. Hay, editors, *Marine Community Ecology*. Sunderland, MA: Sinauer Associates; Bourque, B.J., B.J. Johnson, and R.S. Steneck (2008), Possible prehistoric fishing effects on coastal marine food webs in the Gulf of Maine. p. 165–185 in Rick, T.C., and J.M. Erlandson, editors *Human Impacts on Ancient Marine Ecosystems: A Global Perspective*. Berkeley and Los Angeles: University of California Press.

Page 19

detailed population range estimates based on trade records and nesting beach observations: Jackson (1997); McClenachan, L., J.B.C. Jackson, and M.J.H. Newman (2006), Conservation implications of historic sea turtle nesting beach loss. *Frontiers in Ecology and Environment* 4: 290–296.

yet we have a very informative historical analysis: p. 14–15 in Pauly, D., and J.L. Maclean (2003), *In a Perfect Ocean: The State of Fisheries and Ecosystems in the North Atlantic Ocean*. Washington, DC: Island Press.

On Darwin's birthday in 2009 only 39 percent of Americans believed that evolution occurs: Newport, F. (2009), On Darwin's birthday, only 4 in 10 believe in evolution. *Gallup*, online at: http://www.gallup.com/poll/114544/darwin-birthday-believe-evolution.aspx

Chapter 2

Page 21

The March of Folly: Tuchman, B.W. (1984), *The March of Folly: From Troy to Vietnam*. New York: Alfred A. Knopf.

Page 22

the decline of fish biomass that occurred in the last decades: Myers, R.A., and B. Worm (2003). Rapid worldwide depletion of predatory fish communities. *Nature* 423: 280–283.

intensifying earlier trends initiated hundreds to thousands of years ago: Jackson, J.B.C., M.X. Kirby, W.H. Berger, K.A. Bjorndal, L.W. Botsford, B.J. Bourque, R.H. Bradbury, R. Cooke, j. Erlandson, J.A. Estes, T.P. Hughes, S. Kidwell, C.B. Lange, H.S. Lenihan, J.M.Pandolfi, C.H. Peterson, R.S. Steneck, M.J. Tegner, and R.R. Warner (2001), Historical overfishing and the recent collapse of coastal ecosystems. *Science* 293: 629–637; Rosenberg, A.A, W.J. Bolster, K.E. Alexander, W.B. Leavenworth, A.B. Cooper, and M.G. McKenzie (2005), The history of ocean resources: Modeling cod biomass using historical records. *Frontiers in Ecology and the Environment* 3(2): 84–90.

which affects predominantly the large, slow-growing, and expensive predatory fish on top of marine food webs: Pauly, D., V. Christensen, J. Dalsgaard, R. Froese, and F.C. Torres Jr. (1998), Fishing down marine food webs. *Science* 279: 860–863.

as well as species facing extinction: Dulvy, N.K., Y. Sadovy, and J.D. Reynolds (2003), Extinction vulnerability in marine populations. *Fish and Fisheries* 4: 25–64; Cheung, W.W.L., T.J. Pitcher, and D. Pauly (2004), A fuzzy logic expert system for estimating the intrinsic vulnerabilities of seamount fishes relative to fishing. p. 51–60 in Morato, T., and D. Pauly, editors *Seamounts: Biodiversity and Fisheries*. Vanrouver BC: Fisheries Centre Research Reports 12(5).

low biomasses tend to fluctuate more widely than high biomasses, which are usually composed of numerous and overlapping year classes: Longhurst, A. (1998), Cod: Perhaps if we all stood back a bit? *Fisheries Research* 38: 101–108; Pauly, D., V. Christensen, S. Guénette, T. Pitcher, U.R. Sumaila, C.J. Walters, R. Watson, and D. Zeller (2002), Toward sustainability in world fisheries. *Nature* 418: 689–695.

using a number of historical vignettes: Tuchman (1984).

the definition of biological and economic overfishing that emerged in the mid-1950s: Ricker, W.E. (1954), Stock and recruitment. Journal of the Fisheries Research Board of Canada 11: 559–623; Schaefer, M.B. (1954), Some aspects of the dynamics of populations important to the management of the commercial marine fisheries. *Bulletin of the Inter-American Tropical Tuna Commission* 1: 27–56; Schaefer, M.B. (1957), A study of the dynamics of the fisheries for yellowfin tuna in the eastern tropical Pacific Ocean. *Bulletin of the Inter-American Tropical Tuna Commission* 2: 247–268; Beverton, R.J.H., and S.J. Holt (1957), *On the Dynamics of Exploited Fish Populations. Fisheries Investigations*, Series II, Vol. 19. London: Chapman and Hall; Gordon, H.S. (1954), The economic theory of common property resource: The fishery. *Journal of Political Economy* 62: 124–142.

overexploitation was followed by the collapse of one species after the other, and the subsequent expansion of the fishing fleets into more distant or deeper fishing grounds: Pauly et al. (2002); Pauly, D., J. Alder, E. Bennett, V. Christensen, P. Tyedmers, and R. Watson (2003), The future for fisheries. *Science* 302: 1359–1361.

this was subsidized with public funds: Sumaila, U.R., A.S. Khan, A.J. Dyck, R. Watson, G. Munro, P. Tyedmers, and D. Pauly (2010), A bottom-up reestimation of global fisheries subsidies. *Journal of Bioeconomics* 12: 201–225.

leading to a peak in global catches in the late 1980s with subsequent, ongoing declines: Watson, R., and D. Pauly (2001), Systematic distortions in world fisheries catch trends. *Nature* 414: 534–536.

Page 23

Figure 2.1 Watson, R., and D. Pauly (2001).

but to no avail: Finlayson, A.C. (1994), *Fishing for Truth: A Sociological Analysis of Northern Cod Stock Assessments from 1977 to 1990*. St. John's, Newfoundland: ISER.

Page 24

the way fisheries have been run since time immemorial: Jackson et al. (2001); Lotze, H.K., H.S. Lenihan, B.J. Bourque, R.H. Bradbury, R.G. Cooke, M.C. Kay, S.M. Kidwell, M.X. Kirby, C.H. Peterson, and J.B.C. Jackson (2006), Depletion, degradation, and recovery potential of estuaries and coastal seas. *Science* 312: 1806–1809; Roberts, C. (2007), *The Unnatural History of the Sea*. Washington, D.C.: Island Press; Bolster, W.J. (2008), Putting the ocean in Atlantic history: Maritime communities and marine ecology in the Northwest Atlantic, 1500–1800. *American Historical Review* 113: 19–47.

claiming they never existed: Mowat, F. (1984), *Sea of Slaughter*. Boston and New York: Atlantic Monthly Press; Pauly, D. (1996), One hundred million tonnes of fish, and fisheries research. *Fisheries Research* 25(1): 25–38.

the folly of excessive quota continuing even after a moratorium was declared in 1992: Walters, C.J., and J-J. Maguire (1996), Lessons for stock assessment from the northern cod collapse. *Reviews in Fish Biology and Fisheries* 6(2): 125–137.

Page 25

declined by a factor of 10 within one to twenty years of being accessed by that fishery: Myers and Worm (2003).

for the North Atlantic from 1900 to 2000: Christensen, V., S. Guénette, J.J. Heymans, C.J. Walters, R. Watson, D. Zeller, and D. Pauly (2003), Hundred-year decline of North Atlantic predatory fishes. *Fish and Fisheries* 4: 1–24.

for Southeast Asia from 1960 to 2000: Christensen, V., L. Garces, G.T. Silvestre, and D. Pauly (2003), Fisheries impact on the South China Sea Large Marine Ecosystem: A preliminary analysis using spatially-explicit methodology. p. 51–62 in Silvestre, G.T, L.R. Garces, I. Stobutzki, M. Ahmed, R.A. Valmonte-Santos, C.Z. Luna, L. Lachica-Aliño, P. Munro, V. Christensen, and D. Pauly, editors *Assessment, Management and Future Directions for Coastal Fisheries in Asian Countries*. Penang, Malaysia: WorldFish Center Conference Proceedings 67.

for North West Africa from 1960 to 2000: Christensen, V., P. Amorim, I. Diallo, T. Diouf, S. Guénette, J.J. Heymans, A. Mendy, M. Ould Taleb Ould Sidi, M.-L.D. Palomares, B. Samb, K.A. Stobberup, J.M. Vakily, M. Vasconcellos, R. Watson, and D. Pauly (2004), Trends in fish biomass off northwest Africa, 1960–2000. In Ba, M., P. Chavance, D. Gascuel, M. Vakily, and D. Pauly, editors *Pêcheries maritimes, écosystèmes et sociétés : un demi-siècle de changement*. Paris: Institut de recherché pour le developpement.

necessary for a stock to generate a harvestable surplus yield: Schaefer (1954, 1957).

assembled by the late Ram Myers: Myers, R.A. (Nd), *Ransom Myers' Stock Recruitment Database*, online at: http://ram.biology.dal.ca/~myers/welcome.html

North Atlantic catches peaked in the mid-1970s: p. 29–30 in Pauly, D., and J.L. Maclean (2003), *In a Perfect Ocean: The State of Fisheries and Ecosystems in the North Atlantic Ocean*. Washington, DC: Island Press.

It has now reached its natural limits: Swartz, W., E. Sala, S. Tracey, R. Watson, and D. Pauly (2010), The spatial expansion and ecological footprint of fisheries (1950 to present). *PLoS One* 5(12): e15143 (doi:10.1371/journal.pone.0015143).

"fishing down marine food webs" is far more pervasive than originally estimated: Pauly, D., and M.-L. Palomares (2005), Fishing down marine food web: It is far more pervasive than we thought. *Bulletin of Marine Science* 76(2): 197–211.

and to other areas or regions of the world: Cushing, D.H. (1988), *The Provident Sea*. Cambridge, UK: Cambridge University Press.

Page 26

Figure 2.3 Watson, R., A. Kitchingman, A. Gelchu, and D. Pauly (2004), Mapping global fisheries: Sharpening our focus. *Fish and Fisheries* 5(2): 168–177.

Page 27

distant water fleets from western and eastern Europe, and from Asia between 1960 and 2000: Alder, J., and U.R. Sumaila (2004), Western Africa: A fish basket of Europe past and present. *Journal of Environment and Development* 13: 156–178.

declines in biomass in the waters off West Africa: Christensen et al. (2004).

received no real economic or social benefits for their depleted marine resources: Kaczynski, V.M., and D.L. Fluharty (2002), European policies in West Africa: Who benefits from fisheries agreements? *Marine Policy* 26: 75–93.

as has been shown for Ghana's coastal fishing communities: Iheduru, O.C. (1996), The political economy of Euro-African fishing agreements. *Journal of Developing Areas* 30: 63–90; Atta-Mills, J., J. Alder, and U.R. Sumaila (2004), The decline of a regional fishing nation: The case of Ghana and West Africa. *Natural Resources Forum* 28: 13–21; Alder and Sumaila (2004).

the destruction of marine resources will ultimately hamper the ability of future generations to meet their needs from marine ecosystems: Clark, C.W. (1973), The economics of overexploitation. *Science* 181: 630–634; Sumaila, U.R. (2004), Intergenerational cost-benefit analysis and marine ecosystem restoration. *Fish and Fisheries* 5: 329–343; Sumaila, U.R., and C. Walters (2005), Intergenerational discounting: A new intuitive approach. *Ecological Economics* 52: 135–142; Ainsworth, C.H., and U.R. Sumaila (2005), Intergenerational valuation of fisheries resources can justify long-term conservation: A case study in Atlantic cod (*Gadus morhua*). *Canadian Journal of Fisheries and Aquatic Sciences* 62: 1104–1110.

overcapitalization and overexploitation: Gordon (1954).

problems due to the transboundary or "shared" nature of some fishery resources: Munro, G.R. (1979), The optimal management of transboundary renewable resources. *Canadian Journal of Economics* 12: 355–376; Sumaila, U.R. (1997), Cooperative and non-cooperative exploitation of the Arcto-Norwegian cod stock. *Environmental and Resource Economics* 10: 147–165.

Page 28

which intensify overcapitalization and overfishing: Munro, G., and U.R. Sumaila (2002), The impact of subsidies upon fisheries management and sustainability: The case of the North Atlantic. *Fish and Fisheries* 3: 233–250.

amount of subsidies to the fishing sector is nearly $30 billion per year: Sumaila, U.R., A.S. Khan, A.J. Dyck, R. Watson, G. Munro, P. Tyedmers, and D. Pauly (2010), A bottom-up re-estimation of global fisheries subsidies. *Journal of Bioeconomics* 12: 201–225.

which presently cover only about 0.7 percent of the area of the world's ocean: Wood, L., L. Fish, J. Laughren, and D. Pauly (2008), Assessing progress towards global marine protection targets: Shortfalls in information and action. *Oryx* 42(3): 340–351.

Page 29

compared to net benefits that can be achieved today: Koopmans, T.C. (1960), Stationary ordinal utility and impatience. Econometrica 28: 287–309; Heal, G.M. (1997), Discounting and climate change; an editorial comment. *Climate Change* 37: 335–343.

time stream comparisons can have a huge impact on the apparent best policy or project: Koopmans (1960); Heal (1997).

unsustainable use of natural resources and particularly to global overfishing: Clark (1973); Koopmans, T.C. (1974), Proof of a case where discounting advances doomsday. *Review of Economic Studies* 41: 117–120.

the recent collapse of northern cod off Newfoundland: Ainsworth and Sumaila (2005).

a matter of policy rather than a scientific question: Fearnside, P.M. (2002), Time preference in global warming calculations: A proposal for a unified index. *Ecological Economics* 41: 21–31.

the interests of future generations be taken into account in the management of U.S fisheries: *Magnuson-Stevens Fishery Conservation and Management Act*. U.S. Public Law, 94-265. J. Feder version (12/19/1996 [revised 2006]).

the choice of discount rate and discounting approach is both empirical and ethical: Tol, R.S.J. (1999), Time discounting and optimal emission reduction: An application of FUND. *Climate Change* 41: 351–362.

Page 30

while using a higher rate for their personal decisions: Intergovernmental Panel on Climate Change (2001), *Climate change 2001: The scientific basis. Contribution of Working Group 1 to the Third Assessment Report, Summary for policy makers*, online at: http://www.ipcc.ch; Marglin, S.A. (1963), The social rate of discount and the optimal rate of investment. *Quarterly Journal of Economics* 77: 95–111.

actually care about benefits to generations yet unborn: Popp, D. (2001), Altruism and the demand for environmental quality. *Land Economics* 77(3): 339–350.

overcoming shortsighted decision making would be the right thing to do: Rawls, J. (1971), *A Theory of Justice*. Cambridge, MA: The Belknap Press of Harvard University Press.

compatible with legal and cultural constraints in Western societies: Macinko, S., and D.W. Bromley (2002), *Who Owns America's Fisheries?* Washington, DC: Island Press.

reestablish the natural barriers to fishing that technological progress has removed: Pauly et al. (2002).

Page 31

creating a global network of marine reserves: Russ, G.R., and D. Zeller (2003), From *Mare Liberum* to *Mare Reservarum*. *Marine Policy* 27(1): 75–78; Wood et al. (2008).

provide protection against assessment errors, acknowledged to be a common cause of collapses: Walters, C.J. (1998), Designing fisheries management systems that do not depend upon accurate stock assessment. p. 279–288 in Pitcher, T.J., D. Pauly, and P. Hart, editors *Reinventing Fisheries Management*. London: Chapman and Hall.

overreporting besetting Chinese fisheries catch statistics: Watson, R., and D. Pauly (2001), Systematic distortions in world fisheries catch trends. *Nature* 414: 534–536.

all of which consume more fish products than they themselves contribute: Pauly, D., P. Tyedmers, R. Froese, and L.Y. Liu (2001), Fishing down and farming up the food web. *Conservation Biology in Practice* 2(4): 25; Naylor, R.L., R.J. Goldberg, J.H. Primavera, N. Kautsky, M.C.M. Beveridge, J. Clay, C. Folke, J. Lubchenco, H. Mooney, and M. Troell (2000), Effect of aquaculture on world fish supplies. *Nature* 405: 1017–1024.

the interest of future generations as well as the present generation: Ainsworth and Sumaila (2005); Clark (1973); Sumaila (2004); Sumaila and Walters (2005).

Chapter 3

Page 33

Pew Oceans Commission Report: Pew Oceans Commission (2003), *America's Living Oceans: Charting a Course for Sea Change*, online at: http://www.pewtrustsorg/our_work_report_detail.aspx?id=30009&category=130

Page 34

Shifting Baselines Ocean Media Project: Olson, R., and T. Carlisle (Nd.), *Shifting Baselines Ocean Media Project*, online at: http//www.shiftingbaselines.org.

to assess the condition of America's coastal oceans: Pew Oceans Commission (2003).

the Stratton Commission Report of 1969: Stratton Commission on Marine Science, Engineering and Resources (1969), *Our Nation and the Sea. A Plan for National Action*. Report of the Commission. Washington, DC: U.S. Government Printing Office, online at: http://www.lib.noaa.gov/noaainfo/heritage/stratton/contents.html

the 1969 Santa Barbara oil spill: Easton, R. (1972), *Black Tide: The Santa Barbara Oil Spill and Its Consequences*. New York: Delacorte Press.

Page 36

single ideas, data points, observations, factoids, or reports are what change the world: Olson, R. (2009), *Don't Be Such a Scientist: Talking Substance in an Age of Style*. Washington, DC: Island Press.

instead ended up on page A-22 of that venerable publication: Rivkin, A.C. (2003), U.S. is urged to overhaul its approach to protecting oceans. *New York Times* 06-05-2003, online at: http://www.nytimes.com/2003/06/05/us/us-is-urged-to-overhaul-its-approach-to-protecting-oceans.html?scp=1&sq=Pew%20oceans%20commission&st=cse

the final conclusions of the 9/11 Commission: National Commission on Terrorist Attacks upon the United States (2004), *The 9/11 Commission Report*. New York: W.W. Norton & Company.

Page 37

leaving the report to speak for itself: Shenon, P. (2004), Sept. 11 Commission plans a lobbying campaign to push its recommendations. *New York Times*. 07-19-2004, online at: http://query.nytimes.com/gst/fullpage.html?res=9E0CEFDC133AF93AA25754C0A9629C8B63&scp=4&sq=9/11%20report%20lobbying%20effort&st=cse

Ken Auletta detailed in an article in *The New Yorker* entitled "The New Pitch: Do ads still work?": Auletta, K. (March 28, 2005), The New Pitch: Do ads still work? *The New Yorker*, online at: http//www.newyorker.com/archive/2005/03/28/050328fa_fact.

Page 38

Figure 3.2 illustrates this point. IMDb.com, Inc. (Nd.), *Box Office Figures*, online at: http://www.boxofficemojo.com

Page 39

he recounts the gigantic surge in sales it brought them: Bardley, T. (2008), Super Political Ads Don't Stake Up to Super Bowl Ads. *ABC News/Politics*, aired February 4, online at: http://abcnews.go.com/Politics/story?id=4240705&page=1

Page 40

Turning the Tide: Charting a Course to Improve the Effectiveness of Public Advocacy for the Oceans: Wilmot, D., and J. Sterne (2003), *Turning the Tide Final Report*, online at: http://www.oceanchampions.org/turningthetide.htm

Jennifer Jacquet and her "Guilty Planet" blog: Jacquet, J. (2010), *Guilty Planet*. New York: ScienceBlogs LLC. online at: http://scienceblogs.com/guiltyplanet/

Page 41

Al Gore's movie, *An Inconvenient Truth*, released in 2006: Guggenheim, D., D. Weyerman, J. Skoll, J.D. Ivers, L. Lennard (2006), *An Inconvenient Truth*. Hollywood, CA: Paramount Films.

"Those who cannot remember the past are condemned to repeat it": p. 284, vol. 1 in Santayana, G. (1905), *The Life of Reason; or, the Phases of Human Progress*, New York: Charles Scribner's Sons.

The summary report of the United States Global Change Program: United States Global Change Research Program (2009), *White House Releases Landmark Climate Change Report*. 06-16-2009. online at: http://www.globalchange.gov/publications/reports/scientific-assessments/us-impacts

In a Perfect Ocean: Pauly, D., and J.L. Maclean (2003), *In a Perfect Ocean: The State of Fisheries and Ecosystems in the North Atlantic Ocean*. Washington, DC: Island Press.

Chapter 4

Page 48

implying that fishing pressure need not be reduced: Clark, F.N., and J.C. Marr (1955), Population dynamics of the Pacific sardine. *California Cooperative Oceanic Fisheries Investigations Progress Report* (1 July 1953–31 March 1955): 11–48.

the contemporary ecological debate regarding density dependent or density independent control of animal populations: e.g., Andrewartha, H.G., and L.C. Birch (1954), *The Distribution and Abundance of Animals*. Chicago: University of Chicago Press; Nicholson, A.J. (1957), The self-adjustment of populations to change. *Cold Spring Harbor Symposium on Quantitative Biology* 22: 153–173.

"squeezed out what life remained in the sardine fishery": p. 202 in McEvoy, A.F. (1986), *The Fisherman's Problem: Ecology and Law in California Fisheries 1850–1980*. New York: Cambridge University Press.

Page 49

the now-standard fishery stock assessment tool of Virtual Population Analysis: Murphy, G.I. (1965), A solution of the catch equation. *Journal of the Fisheries Research Board, Canada* 22: 191–202; and Murphy (1966), Population biology of the Pacific sardine (*Sardinops caerulea*). *Proceedings of the California Academy of Science* 34(1): 1–84.

reproductive failure in 1949 and 1950 precipitated the collapse of the stock: Murphy (1966).

Page 50

it was not economically viable: Thomson, C.J. (1990), The market for fish meal and oil in the United States: 1960–1988 and future prospects. *California Cooperative Oceanic Fisheries Investigations Report* 31: 124–131; Jacobson, L.D., and C.J.

Thomson (1993), Opportunity costs and the decision to fish for northern anchovy. *North American Journal of Fisheries Management* 13: 27–34.

The recreational fishing sector was militantly opposed to a large anchovy fishery: Izor, R. (1969), The point of view of the partyboat and live bait industries. *California Cooperative Oceanic Fisheries Investigations Report* 13: 113–116.

spawning biomass in the late 1960s had actually been less than 500,000 mt: Lo, N.C.H., and R.D. Methot (1989), Spawning biomass of the northern anchovy in 1988. *California Cooperative Oceanic Fisheries Investigations Report* 30: 18–31.

proposed annual anchovy harvests of 200,000 to 1,000,000 mt: chapters 9–10 in McEvoy (1986).

a competing source of mortality in the Peruvian anchoveta (*Engraulis ringens*) fishery: Schaefer, M.B. (1970), Men, birds and anchovies in the Peru current-Dynamic interactions. *Transactions of the American Fisheries Society* 99: 461–467.

"the annual anchoveta catch can be maintained indefinitely at 9.3 million metric tons": Schaefer (1970).

Page 51

unusually high vulnerability to an already intense fishery: Csirke, J. (1980), Recruitment in the Peruvian anchovy and its dependence on the adult population. *Rapports et Procès-Verbaux des Réunions, Conseil International pour L'Exploration de la Mer* 177: 307–313.

Sardinops melanostictus was increasing even faster in Japan: Kondo, K. (1980), The recovery of the Japanese sardine-The biological basis of stock-size fluctuations. *Rapports et Procès-Verbaux des Réunions, Conseil International pour L'Exploration de la Mer* 177: 332–354.

fish scales preserved in southern California laminated anaerobic sediments: Soutar, A., and J.D. Isaacs (1969), History of fish populations inferred from fish scales in anaerobic sediments off California. *California Cooperative Oceanic Fisheries Investigations Report* 13: 63–70; Soutar, A., and J.D. Isaacs (1974), Abundance of pelagic fish during the 19th and 20th centuries as recorded in anaerobic sediment off the Californias. *Fishery Bulletin* (U.S.) 72: 257–273.

since refined by Baumgartner and colleagues: Baumgartner, T.R., A. Soutar, and V. Ferreira-Bartrina (1992), Reconstruction of the history of Pacific sardine and northern anchovy populations over the past two millennia from sediments of the Santa Barbara Basin, California. *California Cooperative Oceanic Fisheries Investigations Report* 33: 24–40.

"Nor can the virtual absence of the sardine from the waters off Alta California be considered an unnatural circumstance": Soutar and Isaacs (1974).

at which level zero scale counts become frequent: Lasker, R., and A.D. MacCall (1983), New ideas on the fluctuations of the clupeoid stocks off California. p. 110–120 in *Proceedings of the Joint Oceanographic Assembly 1982—General Symposia*. Ottawa, Canada: Canadian National Committee/Scientific Committee on Oceanic Research.

Figure 4.2: Baumgartner, Soutar and Ferreira-Bartrina (1992).

Page 52

despite scientific consensus that the two species were competitors: e.g., Sette, O.E. (1969), A perspective of a multi-species fishery. *California Cooperative Oceanic Fisheries Investigations Report* 13: 81–87.

"Fluctuations of populations must be related to these very large alternations of conditions": Isaacs, J.D. (1976), Some ideas and frustrations about fishery science. *California Cooperative Oceanic Fisheries Investigations Report* 18: 34–43.

from Soutar and Isaacs' study: Soutar and Isaacs (1974).

the average anchovy was 54 percent heavier during periods when sardine scale deposition was low: Lasker and MacCall (1983).

a sudden reduction in the average size of anchovies had occurred in southern California: Mais, K.F. (1981), Age-composition changes in the anchovy (*Engraulis mordax*) central population. *California Cooperative Oceanic Fisheries Investigations Report* 22: 82–87.

Page 53

Expert Consultation to Examine Changes in Abundance and Species Composition of Neritic Fish Resources: Sharp, G.D., and J. Csirke, editors (1983), *Proceedings of the Expert Consultation to Examine Changes in Abundance and Species Composition of Neritic Fish Resources*. San Jose, Costa Rica (18–29 April): FAO Fisheries Report 291.

a figure presented by Kawasaki, reproduced here in figure 4.3: Kawasaki, T. (1983), Why do some pelagic fishes have wide fluctuations in their numbers?-Biological basis of fluctuation from the viewpoint of evolutionary ecology. p. 1065–1080 in *Proceedings of the Expert Consultation to Examine Changes in Abundance and Species Composition of Neritic Fish Resources*.

the fishery-oceanographic mechanisms governing fish recruitment in eastern boundary currents worldwide: Parrish, R.H., A. Bakun, D.M. Husby, and C.S. Nelson (1983), Comparative climatology of selected environmental processes in relation to eastern boundary current pelagic fish reproduction. p. 731–777 in *Proceedings of the Expert Consultation to Examine Changes in Abundance and Species Composition of Neritic Fish Resources*.

regime shift: sensu Isaacs (1976), though they did not use the term.

Figure 4-3: Kawasaki (1983).

Page 54

"and our studies of community structure and dynamics": Venrick, E.L., J.A. McGowan, D.R. Cayan, and T.L. Hayward (1987), Climate and chlorophyll a: Long-term trends in the Central North Pacific Ocean. *Science* 238: 70–72.

a multivariate study in the meteorological literature by Trenberth in 1990: Trenberth, K.E. (1990), Recent observed interdecadal climate changes in the Northern Hemisphere. *Bulletin of the American Meteorological Society* 71: 988–993.

worldwide decadal scale variability of anchovies and sardines: Lluch-Belda, D.,

R.J.M. Crawford, T. Kawasaki, A.D. MacCall, R.H. Parrish, R.A. Schwartzlose, and P.E. Smith (1989), World-wide fluctuations of sardine and anchovy stocks: The regime problem. *South African Journal of Marine Science* 8: 195–205.

if sardine abundance recovered to at least 20,000 short tons (18,144 mt): Wolf, P. (1992), Recovery of the Pacific sardine and the California sardine fishery. *California Cooperative Oceanic Fisheries Investigations Report* 33: 76–86.

abundance had reached this level and a small fishery was allowed: Wolf, P., and P.E. Smith (1986), The relative magnitude of the 1985 Pacific sardine spawning biomass off southern California. *California Cooperative Oceanic Fisheries Investigations Report* 27: 25–31.

an ichthyoplankton-based spawning area survey in the southern California Bight: Wolf, P., and P.E. Smith (1985), An inverse egg production method for determining the relative magnitude of Pacific sardine spawning biomass off California. *California Cooperative Oceanic Fisheries Investigations Report* 26: 130–138.

the resource biomass declined by 95 percent between 1988 and 1992: Wada, T., and L.D. Jacobson (1998), Regimes and stock-recruitment relationships in Japanese sardine (*Sardinops melanostictus*), 1951–1995. *Canadian Journal of Fisheries and Aquatic Sciences* 55: 2455–2463.

Page 55

anchoveta catches were returning to pre-1972 levels: Chavez, F.P., J. Ryan, S.E. Lluch-Cota, and M. Ñiquen C. (2003), From anchovies to sardines and back: Multidecadal change in the Pacific Ocean. *Science* 299: 217–221.

the southern California spawning area increased progressively from 1985 to 1991: Barnes, J.T., L.D. Jacobson, A.D. MacCall, and P. Wolf (1992), Recent population trends and abundance estimates for the Pacific sardine (*Sardinops sagax*). *California Cooperative Oceanic Fisheries Investigations Report* 33: 60–75.

the farthest edge of the range covered by standard CalCOFI surveys: Macewicz, B.J., and D.N. Abramenkoff (1993), Collection of jack mackerel (*Trachurus symmetricus*) off southern California during 1991 cooperative U.S.- U.S.S.R. cruise. *NMFS SWFSC Administrative Report* LJ-93-07.

after a nearly forty-year absence: Hargreaves, N.B., D.M. Ware, and G.A. McFarlane (1994), Return of Pacific sardine (*Sardinops sagax*) to the British Columbia coast in 1992. *Canadian Journal of Fisheries and Aquatic Sciences* 51: 460–463.

Mantua and colleagues described the Pacific Decadal Oscillation (PDO): Mantua, N.J., S.R. Hare, Y. Zhang, J.M. Wallace, and R.C. Francis (1997), A Pacific interdecadal climate oscillation with impacts on salmon production. *Bulletin of the American Meteorological Society* 78: 1069–1079.

stock and recruitment relationship were developed for both the Pacific sardine: Jacobson, L., and A.D. MacCall (1995), Stock-recruitment models for Pacific sardine (*Sardinops sagax*). *Canadian Journal of Fisheries and Aquatic Sciences* 52: 566–577.

and the Japanese sardine: Wada and Jacobson (1998).

Page 56

recruits per spawner were about twice as high during favorable environmental conditions as they were during unfavorable conditions: Jacobson and MacCall *(1995).*

Japanese sardine achieved a remarkable twentyfold increase in recruitment during favorable environmental conditions: Wada and Jacobson (1998).

low-frequency variability in sardine and anchovy systems around the world: Schwartzlose, R., J. Alheit, A. Bakun, T. Baumgartner, R. Cloete, R. Crawford, W. Fletcher, Y. Green-Ruiz, E. Hagen, T. Kawasaki, D. Lluch-Belda, S. Lluch-Cota, A. MacCall, Y. Matsuura, M. Nevarez-Martinez, R. Parrish, C. Roy, R. Serra, K. Shust, M. Ward, and J. Zuzunaga (1999), Worldwide large-scale fluctuations of sardine and anchovy populations. *South African Journal of Marine Science* 21: 289–347.

"the underlying mechanisms . . . have yet to be identified": McFarlane, G.A., P.E. Smith, T.R. Baumgartner, and J.R. Hunter (2002), Climate variability and Pacific sardine populations and fisheries. *American Fisheries Society Symposium* 32: 195–214.

"they increase off California and Peru when those regions warm and become less productive": Chavez et al. (2003).

Page 57

historical fishery catches and population estimates given by MacCall in 1979: MacCall, A.D. (1979), Population estimates for the waning years of the Pacific sardine fishery. *California Cooperative Oceanic Fisheries Investigations Report* 20:72–82.

which has long been considered to be a safe rule of thumb for fishery management: Alverson, D.L., and W.T. Pereyra (1969), Demersal fish exploration in the north-eastern Pacific Ocean; an evaluation of exploratory fishing methods and analytical approaches to stock size and yield forecasts. *Journal of the Fisheries Research Board, Canada* 26: 1985–2001; Gulland, J.A. (1983), *Fish Stock Assessment: A Manual of Basic Methods*. New York: John Wiley and Sons.

A fishery management plan for California's sardine fishery was adopted in 1998: Bargmann, G., D. Hanan, S.F. Herrick, K. Hill, L. Jacobson, J. Morgan, R., Parrish, J. Spratt, and M. Walker (1998), *The Coastal Pelagic Species Management Plan.* Portland, OR: Pacific Fishery Management Council, 2130 SW Fifth Ave., Suite 224, 97201.

stock and recruitment relationships described by Jacobson and MacCall in 1995: Jacobson and MacCall (1995).

Chapter 5

Page 59

seasonal quotas can be reached shortly after the fishing season opens: Fréon, P., M. Bouchon, C. Mullon, C. García, and M. Ñiquen (2008), Interdecadal variability of anchoveta abundance and overcapacity of the fishery in Peru. *Progress in Oceanography* 79(2–4): 401–412.

in the California Current, Kuroshio Current, Benguela Current, and many other regions: Schwartzlose, R., J. Alheit, A. Bakun, T. Baumgartner, R. Cloete, R. Crawford, W. Fletcher, Y. Green-Ruiz, E. Hagen, T. Kawasaki, D. Lluch-Belda, S. Lluch-Cota, A. MacCall, Y. Matsuura, M. Nevarez-Martinez, R. Parrish, C. Roy, R. Serra, K. Shust, M. Ward, and J. Zuzunaga (1999), Worldwide large-scale fluctuations of sardine and anchovy populations. *South African Journal of Marine Science* 21: 289–347.

Page 60

Landings declined sharply and remained persistently low: MacCall, this volume.

Page 61

Figure 5.2: Fréon et al. (2008); Jahncke, J., D.M. Checkley Jr., and G.L. Hunt Jr. (2004), Trends in carbon flux to seabirds in the Peruvian upwelling system: Effects of wind and fisheries on population regulation. *Fisheries Oceanography* 13: 208-223.

Page 62

The Institute of the Seas of Peru (IMARPE) closely monitors anchovy abundance: Gutiérrez, M., G. Swartzman, A. Bertrand, S. Bertrand (2007), Anchovy (*Engraulis ringens*) and sardine (*Sardinops sagax*) spatial dynamics and aggregation patterns in the Humboldt Current ecosystem, Peru, from 1983–2003. *Fisheries Oceanography* 16: 155–168.

this stabilizes the fishery: Ñiquen, M., and M. Bouchon (2004), Impact of El Niño events on pelagic fisheries in Peruvian waters. *Deep Sea Research II* 51: 563–574.

decreases the quality and quantity of fishmeal: Schaefer, M.B. (1967), Dinámica de la pesquería de la anchoveta (*Engraulis ringens*), en el Perú. *Boletín Instituto del Mar de Perú* 1: 189–304.

could yield greater value per fish: Salvatteci, R., and J. Mendo (2005), Estimation of bio-economic losses caused by the capture of juvenile anchovy (*Engraulis ringens*) in the Peruvian coast. *Revista de Ecología Aplicada UNALM* 4: 113–120.

because it threatened seabirds and guano production: Csirke, J., and A. Gumy (1996), Análisis bioeconómico de la pesquería pelágica peruana dedicada a la producción de harina y aceite de pescado. *Boletín Instituto del Mar de Perú* 15: 22–68.

large-scale changes in Pacific climate that persist for decades at a time: Baumgartner et al. (1992); Schwartzlose et al. (1999); Chavez, F.P., J. Ryan, S.E. Lluch-Cota, and M. Ñiquen C. (2003), From anchovies to sardines and back: Multidecadal change in the Pacific Ocean. *Science* 299: 217–221; MacCall, this volume.

these highly productive, variable, and relatively well studied fisheries: for a full review see Fréon, P., P. Cury, L. Shannon, and C. Roy (2005), Sustainable exploitation of small pelagic fish stocks challenged by environmental and ecosystem changes: A review. *Bulletin of Marine Science* 76: 385–462; Checkley, D.M., C. Roy, J. Alheit, and Y. Oozeki, editors (2009), *Climate Change and Small Pelagic Fish*. Cambridge UK: Cambridge University Press.

Page 63

long before fishing becomes economically unviable: Schertzer K.W., and M.H. Prager (2007), Delay in fishery management: Diminished yield, longer rebuilding, and increased probability of stock collapse. *ICES Journal of Marine Science* 64: 149–159.

shifted focus to the anchoveta fishery in Peru: McEvoy, A.F. (1996), Historical interdependence between ecology, production, and management in California fisheries. p. 45–53 in Bottom, D., G. Reeves, and M. Brookes, editors, *Sustainability Issues for Resource Managers*. USDA Forest Service Tech Rep. PNWGTR-370.

the capacity to extract several times the number of fish that existed in the wild: Fréon et al. (2008).

Figure 5.3: Schertzer and Prager (2007).

Page 64

population growth and the abundance of anchovies failed to recover for decades: Pauly, D. (1987), Managing the Peruvian upwelling ecosystem: A synthesis. p. 325–342 in Pauly, D., and I. Tsukayama, editors *The Peruvian Anchoveta and Its Upwelling Ecosystem: Three Decades of Change*. ICLARM Studies and Reviews 15, Instituto del Mar del Peru (IMARPE), Callao, Peru; Deutsche Gesellschaft fur Technische Zusammenarbeit (GTZ), GmbH, Eschborn, Federal Republic of Germany; and International Center for Living Aquatic Resources Management (ICLARM), Manilla, Philippines; Chavez et al. (2003).

or density independence because environmental factors dominate: Hilborn, R., and C.J. Walters (1992), *Quantitative Fisheries Stock Assessment: Choice, Dynamics and Uncertainty*. New York: Chapman and Hall.

because competition for food is reduced: van der Lingen, C.D., P. Fréon, T.P. Fairweather, and J.J. van der Westhuizen (2006), Density-dependent changes in reproductive parameters and condition of southern Benguela sardine (*Sardinops sagax*). *African Journal of Marine Science* 28: 625–636.

an adverse effect on recruitment in future years: Pauly (1987).

increases the probability of high recruitment when spawning biomass is low: MacCall, A.D. (1979), Population estimates for the waning years of the Pacific sardine fishery. *California Cooperative Oceanic Fisheries Investigations Report* 20: 72–82.

recruitment increases with greater population size and spawning: Hilborn and Walters (1992).

until a threshold is reached and the relationship breaks down: Barrowman, N.J., and R.A. Myers (2000), Still more spawner-recruitment curves: The hockey stick and its generalizations. *Canadian Journal of Fisheries and Aquatic Sciences* 57: 665–676.

density dependence is important only at low population sizes: Nodo, M., and I. Yasuda (2003), Empirical biomass model for the Japanese sardine (*Sardinops melanostictus*) with sea surface temperature in the Kuroshio Extension. *Fisheries Oceanography 12: 1–9.*

The population threshold may also vary with climatic conditions: Jacobson, L., and A.D. MacCall (1995), Stock-recruitment models for Pacific sardine (*Sardinops sagax*). *Canadian Journal of Fisheries and Aquatic Science* 52: 566–577.

Page 65

Figure 5.4: Castillo, R., M. Gutiérrez, S. Peraltilla, and N. Herrera (1998), Biomasa de recursos pesqueros a finales del invierno de 1998 Crucero BIC Humboldt José Olaya, de Paita a Tacna. *Informe Instituto del Mar de Perú* 141: 136-155; Ñiquen and Bouchon (2004).

Page 66

the incorporation of temperature variation as a factor in the regulation of the fishery: Conser, R.J., K.T. Hill, P.R. Crone, N.C.H. Lo, and D. Bergen (2002), *Stock Assessment of Pacific Sardine with Management Recommendations for 2002*. Executive summary submitted to Pacific Fishery Management Council, Portland, OR.

the relationship between temperature and recruitment is still not well understood: McFarlane, G.A., P.E. Smith, T.R. Baumgartner, and J.R. Hunter (2002), Climate variability and Pacific sardine populations and fisheries. *American Fisheries Society Symposium* 32: 195–214.

emphasized variations in productivity: Chavez et al. (2003).

mechanisms of larval recruitment related to retention and circulation: Logerwell, E., and P. Smith (2001), Mesoscale eddies and survival of late stage Pacific sardine (*Sardinops sagax*) larvae. *Fisheries Oceanography* 10: 13–25.

emphasized flow patterns on basin scales: MacCall, this volume.

sardines were more abundant than anchovies: Chavez et al. (2003).

a weak relationship between zooplankton biomass and fish larvae from 1950 to 1970: McGowan, J.A., D.B. Chelton, and A. Conversi (1996), Plankton patterns, climate, and change in the California Current. *California Cooperative Ocean Fisheries Investigations Report* 37: 45–68.

sardines increased during the 1980s and 1990s when zooplankton decreased: Smith, P.E., and H.G. Moser (2003), Long-term trends and variability in the larvae of Pacific sardine and associated fish species of the California Current region. *Deep Sea Research* II 50: 2519–2536.

which are generally of little dietary importance to fish: Lavaniegos, B.E., and M.D. Ohman (2003), Long-term changes in pelagic tunicates in the California Current. *Deep Sea Research II* 50: 2473–2498.

copepods and euphausiids did not decrease: Rebstock, G.A. (2001), Longterm stability of species composition in calanoid copepods off southern California. *Marine Ecology Progress Series*215: 213–224; Brinton, E., and A. Townsend (2003), Decadal variability in abundances of the dominant euphausiid species in southern sectors of the California Current. *Deep Sea Research II* 50: 2449–2472.

Page 67

complicates stock assessments and understanding of recruitment dynamics: McFarlane et al. (2002).

the Kuroshio Current, California Current, and Peru-Chile Current: Schwartzlose et al. (1999); Fréon et al. (2005).

move south toward Chilean waters: Ñiquen and Bouchon (2004); Gutiérrez et al. (2007).

fish can increase near the coast during a strong El Niño: Csirke, J. (1989), Changes in the catchability coefficient in the Peruvian anchoveta (Engraulis ringens) fishery. p. 207–219 in Pauly, D., P. Muck, J. Mendo, and I. Tsukayama, editors *The Peruvian Upwelling Ecosystem: Dynamics and Interactions*. Instituto del Mar del Perú (IMARPE), Callao, Peru; Deutsche Gesellschaft fur Technische Zusammenarbeit (GTZ), GmbH, Eschborn, Federal Republic of Germany; and International Center for Living Aquatic Resources Management (ICLARM), Manila, Philippines.

but migrations occur between them: Pauly (1987).

show a more wide-ranging swimming behavior: McFarlane et al. (2002).

may be due to the existence of a separate stock: McFarlane et al. (2002).

important to commercial and recreational fisheries: Fréon et al. (2005).

Page 68

change migrational patterns to feed in regions with greater availability of prey: Polovina, J.J. (1996), Decadal variation in the trans-Pacific migration of northern bluefin tuna (*Thunnus thynnus*) coherent with climate-induced change in prey abundance. *Fisheries Oceanography* 5: 114–119.

to maintain availability of small pelagics as food for these species: Smith, P.E., personal observation.

prior to the inception of the fishery: Muck, P., and H. Fuentes (1987), Sea lion and fur seal predation on the Peruvian anchoveta, 1953 to 1982. p. 234–247 in Pauly, D., and I. Tsukayama, editors, *The Peruvian Anchoveta and Its Upwelling Ecosystem: Three Decades of Change*. ICLARM Studies and Reviews 15, Instituto del Mar del Peru (IMARPE), Callao, Peru; Deutsche Gesellschaft fur Technische Zusammenarbeit (GTZ), GmbH, Eschborn, Federal Republic of Germany; and International Center for Living Aquatic Resources Management (ICLARM), Manilla, Philippines.

and has not recovered along with the anchoveta: Jahncke, Checkley Jr., and Hunt Jr. (2004).

variability in population size prior to catch records and oceanographic surveys: Soutar, A., and J.D. Isaacs (1974), Abundance of pelagic fish during the 19th and 20th-centuries as recorded in anaerobic sediment off the Californias. *Fishery Bulletin* (U.S.) 72: 257–273; Baumgartner, T.R., A. Soutar, and V. Ferreira-Bartrina (1992), Reconstruction of the history of Pacific sardine and northern anchovy populations over the past two millennia from sediments of the Santa Barbara Basin, California. *California Cooperative Oceanic Fisheries Investigations Report* 33: 24–40; MacCall, this volume.

which seems to be well founded: Shackleton, L.Y. (1988), Scale shedding: An important factor in fossil fish scale studies. *ICES Journal of Marine Science* 44: 259–263; Field, D.B., T.R. Baumgartner, V. Ferreira, D. Gutierrez, H. Lozano-Montes, R. Salvatteci, and A. Soutar (2009), Variability in small pelagic fishes from scales in marine sediments and other historical records. in Checkley, D.M., C. Roy, J. Alheit, and Y. Oozeki, editors, *Climate Change and Small Pelagic Fish*. Cambridge UK: Cambridge University Press.

Page 69

much has been learned from sedimentary records: Field et al. (2009).

unexpectedly appeared in the sardine catch in the late 1980s: Ahlstrom, E.H. (1967), Co-occurrences of sardine and anchovy larvae in the California Cur-rent region off California and Baja California. *California Cooperative Ocean Fisheries Investigations Report* 11: 117–135.

part of the natural variability of the ecosystem: Holmgren-Urba, D., and T.R. Baumgartner (1993), A 250-year history of pelagic fish abundance from anaerobic sediments of the central Gulf of California. *California Cooperative Ocean Fisheries Investigations Report* 34: 60–68.

two thousand years before the advent of commercial fishing: Soutar, A., and J.D. Isaacs (1969), History of fish populations inferred from fish scales in anaerobic sediments off California. *California Cooperative Oceanic Fisheries Investigations Report* 13: 63–70; Baumgartner et al. (1992).

vary out of phase at other timescales, generally on decadal timescales: Baumgartner et al. (1992).

competition or climatic conditions that always favor one species over another: Smith, P.E. (1978), Biological effects of ocean variability: Time and space scales of biological response. *Journal of the International Council for the Exploration of the Sea* 173: 117–127.

a reduction in average size and weight of anchovies: Lasker, R., and A.D. MacCall (1983), New ideas on the fluctuations of the clupeoid stocks off California. p. 110–120 in *Proceedings of the Joint Oceanographic Assembly 1982-General Symposia*. Canadian National Committee/Scientific Committee on Oceanic Research, Ottawa, Canada.

extreme levels during the 1997–98 El Niño: Roemmich, D., and J. McGowan (1995), Climatic warming and the decline of zooplankton in the California

Current. *Science* 267: 1324–1326; Lavaniegos, B.E., and M.D. Ohman (2003), Long-term changes in pelagic tunicates in the California Current. *Deep Sea Research II* 50: 2473–2498.

Page 70

temperate to polar species showed no trend or decrease: Field, D.B., T.R. Baumgartner, C.D. Charles. V. Ferreira-Bartrina, and M.D. Ohman (2006), Planktonic foraminifera of the California Current reflect 20th-century warming. *Science* 311: 63–66.

Figure 5.5: Field et al. (2006).

Page 71

exceeded prior variations throughout the last 150–1,000 years: Urban, F.E., J.E. Cole, and J.T. Overpeck (2000), Influence of mean climate change on climate variability from a 155-year tropical Pacific coral record. *Nature* 407: 989–993; Cobb, K.M., C.D. Charles, H. Cheng, and R.L Edwards (2003), El Niño–Southern Oscillation and tropical Pacific climate during the last millennium. *Nature* 424: 271–276; Charles, C.D., K. Cobb, M.D. Moore, and R.G. Fairbanks (2003), Monsoon-tropical ocean interaction in a network of coral records spanning the 20th century. *Marine Geology* 201: 207–222.

physical and biological shifts in the North Pacific in the mid-1970s: Lange, C.B., S.K. Burke, and W.H. Berger (1990), Biological production off southern California is linked to climatic change. *Climatic Change* 16: 319–329; Miller, A.J., D.R. Cayan, T.P. Barnett, N.E. Graham, and J.M. Oberhuber (1994), The 1976–77 climate shift of the Pacific Ocean. *Oceanography* 7: 21–26; Graham, N. (1995), Simulation of recent global temperature trends. *Science* 267: 666–671.

noteworthy in the late twentieth century and will continue in the future: Mann, M.E., R.S. Bradley, and M.K. Hughes (1999), Northern Hemisphere temperatures during the past millennium: Inferences, uncertainties, and limitations. *Geophysical Research Letters* 26: 759–762; Crowley, T.J. (2000), Causes of climate change over the past 1000 years. *Science* 289: 270–276; Levitus, S., J.I. Antonov, T.P. Boyer, and C. Stephens (2000), Warming of the world ocean. *Science* 287: 2225–2229; Intergovernmental Panel on Climate Change, S. Solomon, D. Qin, and M. Manning, editors (2007), *Climate Change 2007: The Physical Science Basis. Contribution of Working Group I to the Fourth Assessment Report of the Intergovernmental Panel on Climate Change*. Cambridge UK: Cambridge University Press.

subsequent cascades to lower levels: Jackson, J.B.C., M.X. Kirby, W.H. Berger, K.A. Bjorndal, L.W. Botsford, B.J. Bourque, R.H. Bradbury, R. Cooke, J. Erlandson, J.A. Estes, T.P. Hughes, S. Kidwell, C.B. Lange, H.S. Lenihan, J.M. Pandolfi, C.H. Peterson, R.S. Steneck, M.J. Tegner, and R.R. Warner (2001), Historical overfishing and the recent collapse of coastal ecosystems. *Science* 293: 629–637.

commercially extinct by the late nineteenth or early twentieth century: Field, J.C., R.C. Francis, and A. Strom (2001), Toward a fisheries ecosystem plan for the

Northern California Current. *California Cooperative Ocean Fisheries Investigations Report* 42: 74–87.

moving to even more unknown and perhaps unknowable states: Hsieh, C., S.M. Glaser, A.J. Lucas, and G. Sugihara (2005), Distinguishing random environmental fluctuations from ecological catastrophes for the North Pacific Ocean. *Nature* 435: 336–340.

have not been well sampled compared to eggs: Smith, P.E., N.C. Lo, and J. Butler (1992), Life-stage duration and survival parameters as related to interdecadal population variability in Pacific sardine. *California Cooperative Ocean Fisheries Investigations Report* 33: 41–49.

Page 72

sedimentary records from sites off Peru and Chile: Gutiérrez, D., A. Sifeddine, J.L. Reyss, G. Vargas, F. Velazco, R. Salvatteci, V. Ferreira, L. Ortlieb, D. Field, T. Baumgartner, M. Boussafir, H. Boucher, J. Valdés, L. Marinovic, P. Soler, and P. Tapia (2006), Anoxic sediments off Central Peru as a record of interannual to multidecadal changes of climate and upwelling ecosystem during the last two centuries. *Advances in Geosciences* 6: 119–125; Vargas, G., S. Pantoja, J. Rutllant, C. Lange, and L. Ortlieb (2007), Enhancement of coastal upwelling and interdecadal ENSO-like variability in the Peru-Chile Current since late 19th century. *Geophysical Research Letters* 34, L13607, doi:10.1029/2006GL028812; Valdés, J., L. Ortlieb, D. Gutiérrez, L. Marinovic, G. Vargas, and A. Sifeddine (2008), 250 years of sedimentary record of sardine and anchovy scale deposition in Mejillones Bay, Northern Chile. *Progress in Oceanography* 79(2–4): 198–207.

discovery of laminated sediments at Effingham Inlet: Holmgren, D. (2001), *Decadal-Centennial Variability in Marine Ecosystems of the Northeast Pacific Ocean: The Use of Fish Scale Deposition in Sediments*. Seattle, WA: University of Seattle.

population distribution and abundance preceding commercial catch records: Field et al. (2009).

specific links between marine populations and climatic changes: Hsieh et al. (2005).

prevented action from being taken: Shertzer and Prager (2007).

Page 73

within and among different oceanographic regimes: Isaacs, J.D (1976), Some ideas and frustrations about fishery science. *California Cooperative Oceanic Fisheries Investigations Report* 18: 34–43.

add whole new dimensions of uncertainty to the mix: Anderson, C.N.K., C. Hsieh, S.A. Sandin, R. Hewitt, A. Hollowed, J. Beddington, R.M. May, and G. Sugihara (2008), Why fishing magnifies fluctuations in fish abundance. *Nature* 452: 835–839.

although new combinations of species may occur: Field et al. (2006).

coupled with other anthropogenic activities and stresses: Jackson et al. (2001).

severe population bottlenecks or founder events in the recent past: Lecomte, F., W.S. Grant, J.J. Dodson, R. Rodríguez-Sánchez, and B.W. Bowen (2004), Living with uncertainty: Genetic imprints of climate shifts in East Pacific anchovy (Engraulis mordax) and sardine (*Sardinops sagax*). *Molecular Ecology* 13: 2169–2182.

Page 74

Adaptability is all the more important in the face of climate change: Hsieh et al. (2005).

Page 75

will almost certainly have unexpected and unwanted consequences: Yodzis, P. (1988), The indeterminacy of ecological interactions as perceived through perturbation experiments. *Ecology* 69: 508–515; Paine, R.T., M.J. Tegner, and E.A. Johnson (1998), Compounded perturbations yield ecological surprises. *Ecosystems* 1: 535–545; Scheffer, M., S. Carpenter, J.A. Foley, C. Folke, and B. Walker (2001), Catastrophic shifts in ecosystems. *Nature* 413: 591–596.

Part III Introduction

Page 78

scientific results: Rosenberg, A.A, W.J. Bolster, K.E. Alexander, W.B. Leavenworth, A.B. Cooper, and M.G. McKenzie (2005), The history of ocean resources: Modeling cod biomass using historical records. *Frontiers in Ecology and the Environment* 3(2): 84–90.

historical analysis: Leavenworth, W.B. (2006), Opening Pandora's Box: Tradition, competition and technology on the Scotian Shelf, 1852–1860. p. 29–49 in Starkey, D.J., and J.E. Candow, editors, *Studia Atlantica: Proceedings of the 7th Conference of the North Atlantic Fisheries History Association [NAFHA]*. Hull UK and Nordurslod IS: NAFHA.

more than 90 percent from historical levels of abundance: among many others, see Jackson, J.B.C., M.X. Kirby, W.H. Berger, K.A. Bjorndal, L.W. Botsford, B.J. Bourque, R.H. Bradbury, R. Cooke, J. Erlandson, J.A. Estes, T.P. Hughes, S. Kidwell, C.B. Lange, H.S. Lenihan, J.M. Pandolfi, C.H. Peterson, R.S. Steneck, M.J. Tegner, and R.R. Warner (2001), Historical overfishing and the recent collapse of coastal ecosystems. *Science* 293: 629–637; Christensen, V., S. Guénette, J.J. Heymans, C.J. Walters, R. Watson, D. Zeller, and D. Pauly (2003), Hundred-year decline of North Atlantic predatory fishes. *Fish and Fisheries* 4: 1–24; Myers, R.A., and Worm, B. (2003), Rapid worldwide depletion of predatory fish communities. *Nature* 423: 280–283; Lotze, H.K., and I. Milewski (2004), Two centuries of multiple human impacts and successive changes in a North Atlantic food web. *Ecological Applications* 14(5): 1428–1447; Kirby, M.X. (2004), Fishing down the coast: Historical expansion and collapse of oyster fisheries along continental margins. *Proceedings of the National Academy of Sciences of the United States of America* 101(35): 13096–13099; McClenachan, L., J.B.C. Jackson, and M.J.H. Newman (2006), Conservation implications of historic sea turtle nesting beach loss. *Frontiers in Ecology and Environment* 4: 290–296; McClenachan, L., and A.B. Cooper (2008), Extinction rate, historical population structure and ecological role of the Caribbean monk seal. *Proceedings of the Royal Society* B 275: 1351–1358; Limburg, K.E., and J.R. Waldman (2009), Dramatic declines in North Atlantic diadromous fishes. *Bioscience* 59(11): 955–965; Alexander, K.E., W.B. Leavenworth, J. Cournane, A.B. Cooper, S. Claesson, S. Brennan, G. Smith, L. Rains, K. Magness, R. Dunn, T.K. Law, R. Gee, W.J. Bolster, and A.A. Rosenberg (2009), Gulf of Maine cod in 1861: Historical analysis of fishery logbooks, with ecosystem implications. *Fish and Fisheries* 10: 428–449; Hall, C.J., A. Jordaan, and M.G. Frisk (2011), The historic influence of dams on diadromous fish habitat with a focus on river herring and hydrologic longitudinal connectivity. *Landscape Ecology* 26: 95–107.

Chapter 6

Page 79

this phenomenon has deep roots, and deeper implications: Jackson, J.B.C., M.X. Kirby, W.H. Berger, K.A. Bjorndal, L.W. Botsford, B.J. Bourque, R.H. Bradbury, R. Cooke, J. Erlandson, J.A. Estes, T.P. Hughes, S. Kidwell, C.B. Lange, H.S. Lenihan, J.M. Pandolfi, C.H. Peterson, R.S. Steneck, M.J. Tegner, and R.R. Warner (2001), Historical overfishing and the recent collapse of coastal ecosystems. *Science* 293: 629–637; Holm, P., T.D. Smith, and D.J. Starkey, editors (2001), *The Exploited Seas: New Directions for Marine Environmental History. Research in Maritime History* No. 21, International Maritime Economic History Association/Census of Marine Life, St John, Newfoundland; Steneck, R.S., and J.T. Carlton (2001), "Human alterations of marine communities: students beware!" p. 445–468 in Bertness, M.D., S.D. Gaines, and M.E. Hay, editors *Marine Community Ecology*. Sunderland, MA: Sinauer Associates.

Page 80

a paper previously published by our interdisciplinary group: Rosenberg, A.A, W.J. Bolster, K.E. Alexander, W.B. Leavenworth, A.B. Cooper, and M.G. McKenzie (2005), The history of ocean resources: Modeling cod biomass using historical records. *Frontiers in Ecology and the Environment* 3(2): 84–90.

overfishing was identified as an *economic* problem in Europe and the United States during the mid-nineteenth century: Cushing, D.H. (1988), *The Provident Sea*. Cambridge UK: Cambridge University Press.

Page 81

a government inquiry and the foundation of fisheries science: Smith, T.D. (1994), *Scaling Fisheries: The Science of Measuring the Effects of Fishing, 1855–1955*. Cambridge UK: Cambridge University Press.

but was a precondition to it: Rivinus, E.F., and E.M. Youssef (1992), *Spencer Baird of the Smithsonian*. Washington, DC: Smithsonian Institution Press.

convert those seemingly inexhaustible stocks into profitable commodities: Smith, J. (1616), *Description of New England*. London.

producing over 6 million pounds of dried salt cod annually: p. 100, ch. 4 in Vickers, D. (1994), *Farmers and Fishermen: Two Centuries of Work in Essex County, Massachusetts, 1630–1850*. Chapel Hill: University of North Carolina Press.

Madeira, the Azores, and the Canaries, and the Caribbean: data from part 2 in McCusker, J.J., and R.R. Menard (1985), *The Economy of British America, 1607–1789*. Chapel Hill: University of North Carolina Press.

populations of right whales, great auks, sea minks, and various birds: among many sources over four centuries, see: Josselyn, J. (1675; reprint 1865), *An Account of Two Voyages to New England*. William Veazie, Boston; Audubon, J.J., M. Audubon, editor (1897; reprinted 1972), *Audubon and His Journals*. Magnolia, MA: Peter Smith Publishers Inc.; Godin, A.J. (1977), *Wild Mammals of New England*. Baltimore, MD: Johns Hopkins University Press; Starbuck, A. (1878; reprint 1989), *History of the American Whale Fishery*. Secaucus, NJ: Castle Books; Mowat, F. (1984), *Sea of Slaughter*. Boston and New York: Atlantic Monthly Press; Bennett, D.B. (1996), *The Forgotten Nature of New England*. Camden, ME: Downeast Books; Reeves, R.R., J.M. Breiwick, and E.D. Mitchell (1999), History of whaling and estimated kill of right whales (Balaena glacialis) in the northeastern United States, 1620–1924. *Marine Fisheries Review* 61: 1–36; Appolonio, S. (2002), *Hierarchical Perspectives on Marine Complexities: Searching for Systems in the Gulf of Maine*. New York: Columbia University Press; Grasso, G. (2008), What appeared limitless plenty: The rise and fall of the nineteenth-century Atlantic halibut fishery. *Environmental History* 13(1): 66–91.

early Puritan fisheries had perceptible impact on cod stocks south of Cape Ann: Leavenworth, W.B. (2008), The changing landscape of maritime resources in seventeenth-century New England. *International Journal of Maritime History* 20(1): 33–62.

"the probable cause of the rapid diminution of the supply of food-fishes on the coast of New England": quoted on p. 233 in Judd, R.W. (1997), *Common Lands, Common People: The Origins of Conservation in Northern New England*. Cambridge, MA: Harvard University Press.

could reduce valuable populations of pollock, cod, and haddock that entered coastal waters to feed: Connor, S. (1878), *Address*. Augusta: Sprague, Owen and Nash.

Page 82

Figure 6.1: Smith, J. (1635) Map of New England, in *Description of New England*. London.

a U.S. Fish Commission map located 44 pound nets and 150 weirs actively harvesting "river fish": United States Commission of Fish and Fisheries (1873), *Map of Weir and Pound Net Fisheries of Penobscot Bay and River*. Washington, DC: N. Peters.

Page 83

Few reliable catch statistics existed prior to 1900: Murawski, S.A., R.W. Brown, S.X. Cadrin, R.K. Mayo, L. O'Brien, W.J. Overholz, and K.A. Sosebee (1999), *New England groundfish. Our Living Oceans: Report on the Status of U.S. Living Marine Resources, NOAA Technical Memorandum* NMFS-F/SPO-41. U.S. Government Printing Office, Washington, DC; National Oceanic and Atmospheric Administration (NOAA) (October, 1990), *Historical Catch Statistics Atlantic and Gulf Coast States, 1879–1989*. Silver Spring, MD: NOAA.

". . . which could be had from no other source": p. 3 in Bigelow, H.B., and W.C. Schroeder (1953), *Fishes of the Gulf of Maine*. Washington, DC: U.S. Government Printing Office.

data were messy and collection methods unreliable: Worster, D. (1990), The ecology of order and chaos. *Environmental History Review* 14: 1–18.

"shifting baseline syndrome of fisheries": Pauly, D. (1995), Anecdotes and the shifting baseline syndrome of fisheries. *Trends in Ecology and Evolution* 10: 430.

Page 84

the shifting baselines syndrome: Pauly (1995).

the perversion of standards by which the "natural" ocean had come to be evaluated: Jackson et al. (2001); Knowlton, N., and J.B.C. Jackson (2008), Shifting baselines, local impacts, and global change on coral reefs. *PloS Biology* 6: e54.(doi:10.1371/journal.pbio.0060054).

and laboriously dredged up shellfish: Jackson, J.B.C. (1997), Reefs since Columbus. *Coral Reefs* 16 (supplement): S23–S32; Jackson, J.B.C., and E. Sala (2001), Unnatural oceans. *Scientia Marina* 65(Supplement 2): 273–281; Lotze, H.K., H.S. Lenihan, B.J. Bourque, R. Bradbury, R.G. Cooke, M.C. Kay, S.M. Kidwell, M.X. Kirby, C.H. Peterson, and J.B.C. Jackson (2006), Depletion, degradation, and recovery potential of estuaries and coastal seas. *Science* 312: 1806–1809; Rick, T.C., and J.M. Erlandson, editors (2008), *Human Impacts on Ancient Marine Ecosystems: A Global Perspective*. Berkeley: University of California Press.

hunted so heavily that local coastal ecosystems were affected: Domning, D.P. (1972), Steller's sea cow and the origin of North Pacific aboriginal whaling. *Syesis* 5: 187–189; Simenstad, C.A., J.A. Estes, and K.W. Kenyon (1978), Aleuts, sea otters, and alternate stable-state communities. *Science* 200: 403–411; Estes, J.A., D.O. Duggins, and G.B. Rathbun (1989), The ecology of extinctions in kelp forest communities. *Conservation Biology* 3: 252–264.

and the concomitant loss of certain trophic levels: Pandolfi, J.M., R.H. Bradbury, E. Sala, T.P. Hughes, K.A. Bjorndal, R.G. Cooke, D. McArdle, L. McClenachan, M.J.H. Newman, G. Paredes, R.R. Warner, and J.B.C. Jackson (2003), Global trajectories of the long-term decline of coral reef ecosystems. *Science* 301: 955–958.

"fishing down the food web" had been abetted by the shifting baseline syndrome: Pauly, D., V. Christensen, J. Dalsgaard, R. Froese, and F.C. Torres Jr. (1998),

Fishing down marine food webs. *Science* 279: 860–863; Pauly, D., and M.-L. Palomares (2005), Fishing down marine food web: It is far more pervasive than we thought. *Bulletin of Marine Science* 76(2): 197–211; Lotze et al. (2006).

Donald Worster, William Cronon, and Richard White in the United States, and Richard Hoffmann in Canada: Crosby, A.W. (1972), *The Columbian Exchange: Biological and Cultural Consequences of 1492*. Westport, CT: Greenwood Pub. Co.; Worster, D. (1979), *Dust Bowl: The Southern Plains in the 1930s*. New York: Oxford University Press; Cronon, W. (1983), *Changes in the Land: Indians, Colonists, and the Ecology of New England*. New York: Hill and Wang; White, R. (1985), American environmental history: The development of a new historical field. *Pacific Historical Review* 54: 297–335; White, R. (1990), Environmental history, ecology, and meaning. *Journal of American History* 76: 1111–1116; Hoffmann, R.C. (1996), Economic development and aquatic ecosystems in medieval Europe. *The American Historical Review* 101: 631–669; Hoffmann, R.C. (2001), Frontier foods for late medieval consumers: Culture, economy, ecology. *Environment and History* 7: 131–167.

Page 85

as effectively as a fistful of water: Lowenthal, D. (1985), *The Past Is a Foreign Country*. New York: Cambridge University Press; and (1996), *Possessed by the Past: The Heritage Crusade and the Spoils of History*. New York: The Free Press.

History is something different: Anderson, K. (2006), Does history count? *Endeavor* 30(4): 150–155.

terms rather than recasting them as people "just like us": Bolster, W.J. (2006), Opportunities in Marine Environmental History. *Environmental History* 11: 567–597; Bolster, W.J. (2008), Putting the ocean in Atlantic history: Maritime communities and marine ecology in the Northwest Atlantic, 1500–1800. *American Historical Review* 113: 19–47.

"not only with the net but in baskets let down with a stone": quoted on p. 48 in Kurlansky, M. (1997), *Cod: A Biography of the Fish That Changed the World*. New York: Walker and Co.

"become more sociological": Carr, E.H. (1961), *What Is History?* New York: Vintage.

Page 86

Time on the Cross: Fogel, R.W., and S.L. Engerman (1974), *Time on the Cross: The Economics of American Negro Slavery*. Boston: Little, Brown.

Without Consent or Contract: Fogel, R.W. (1989), *Without Consent or Contract: The Rise and Fall of American Slavery*. New York: W.W. Norton.

in other public and private collections: Archives with significant fishing collections in Massachusetts include: National Archives Regional Administration, Waltham [NARA]; Peabody Essex Museum, Salem [PEM]; Marblehead Historical Society, Marblehead; Baker Business Library, Harvard University, Cambridge; Beverly Historical Society, Beverly. In Maine, collections include: Penobscot Bay Marine Museum, Searsport; and Old Berwick Historical Society, South Berwick. The

Smithsonian Library, Washington, D.C.; and the G. Blunt-White Library, Mystic Seaport Museum, Mystic, Connecticut, also house fisheries records, and some documents are scattered among local historical societies, museums, libraries, and private collections. Because manuscripts are cataloged differently in different archives, citations below are idiosyncratic. Archivists who know the collection are indispensable and should be consulted from the first. A full archival citation will be given for fishing logs quoted herein.

Page 87

TORPEDO: TORPEDO log, Captain Larkin West, 1852. (NARA Waltham RG36, Box 88: F530a).

IODINE: IODINE log, Captain Thomas Boden, 1856. (NARA Waltham RG36, Box 71: F486b).

PETREL: PETREL log, Captain Calvin Foster, 1854. (NARA Waltham RG36, Box 81: F512c).

and a true vocational scientist: McKenzie, M.G. (2004), Salem as Athenaeum. p. 91–105 in D. Morrison and N. Schultz, editors, *Salem: Place, Myth and Memory*. Boston: Northeastern University Press.

Page 88

Figure 6.2: DOVE log, Captain John Woodbury, 1852 (NARA Waltham RG36, Box 62, F467a).

HENRY: HENRY log, Captain Simeon Beckford, 1855. (NARA Waltham RG36, Box 69: F482d).

Page 89

legislative requirements for bureaucratic oversight between 1852 and 1866: p. 158–169 in Sabine, L. (1853). *Report on the Principal Fisheries of the American Seas, Prepared for the Treasury Department of the United States*. Washington, DC: Robert Armstrong, Printer; ch. 2 in O'Leary, W. (1996), *Maine Sea Fisheries: The Rise and Fall of a Native Industry, 1830–1890*. Boston: Northeastern University Press.

disastrous state following the American Revolution: p. 149–158 in Sabine (1853).

whether they fished inshore or on the deepwater banks: Innis, H. (1954), *The Cod Fisheries: The History of an International Fishery*. Toronto: University of Toronto Press; O'Leary (1996).

the Collector of Customs in the vessel's home port: p. 169 in Sabine (1853).

no longer interested in subsidizing the New England fishery: ch. 2 in O'Leary (1996).

as the fraction of total fish he had caught: p. 166 in Sabine (1853).

Page 90

SARAH: SARAH log, Captain Charles Trask, 1852. (NARA Waltham RG36, Box 91: F537).

FRANKLIN: FRANKLIN log, Captain Samuel Nelson, 1852. (NARA Waltham RG36, Box 67: F474i).

E. W. FORREST: E. W. FORREST log, Captain Solomon Woodbury, 1852. (NARA Waltham RG36, Box 63: F468a).

ESSEX: ESSEX log, Captain Abraham Trowt Jr., 1852. (NARA Waltham RG36 Box 64: F471a).

HENRY: HENRY log, Captain Elisha Pride, 1852. (NARA Waltham RG36, Box 69: F482b).

what had traditionally functioned, at least from the fishermen's perspective, as a maritime common: Leavenworth, W.B. (2006), Opening Pandora's Box: Tradition, competition and technology on the Scotian Shelf, 1852–1860. p. 29–49 in Starkey, D.J., and J.E. Candow, editors, *Studia Atlantica: Proceedings of the 7th Conference of the North Atlantic Fisheries History Association [NAFHA]* . Hull UK and Nordurslod IS: NAFHA.

Page 91

BELLE: BELLE log, Captain Benjamin Gentlee, 1854. (NARA Waltham RG36, Box 56: F452b).

J. PRINCE: J. PRINCE log, Captain George Elliot, 1852. (NARA Waltham RG36, Box 72: F490a).

LODI: LODI log, Captain Oren Eldridge, 1852. (NARA Waltham RG36, Box 74: F496c).

Page 92

were also the most complete between 1852 and 1859: Sources of data for the Beverly Scotian Shelf fleet; the Newburyport and Frenchman's Bay fleets are as follows: NARA Waltham Record Group [RG] 36, Customs District Records: Salem/Beverly MA District logs-Record #441A-538; fishing agreements-Record #9-13; Frenchman's Bay ME District logs and agreements Record #104, 105. James Duncan Phillips Library, Peabody Essex Museum: Newburyport MA logs-Manuscript Collection [Mss] 282, Box 31.

In the 1850s the Scotian Shelf consisted of Browns Bank, LeHave Bank, Sable Island Bank, Middle Bank, and Banquereau: Garcia, S., and T. Farmer, editors (2000), Large Marine Ecosystems. *U.N. Atlas of the Oceans*, online at: http://www.oceansatlas.org/servlet/CDSServlet?status=ND0zNDQwJmN0bl9pbmZvX3ZpZXdfc2l6ZT1jdG5faW5mb192aWV3X2Z1bGwmNj1lbiYzMz0qJjM3PWtvcw~; Garcia, S., and T. Farmer, editors (2000), LME #8: Scotian Shelf. *Large Marine Ecosystems of the World*, online at: http://www.edc.uri.edu/lme/Text/scotian-shelf.htm

conform to the boundary dividing U.S. and Canadian territorial waters: Halliday, R.G., and A.T. Pinhorn (1990), The delimitation of fishing areas in the Northwest Atlantic. *Journal of Northwest Fishery Science* 10: 1–51.

Page 93

Figure 6.3: Garcia and Farmer (2000), Large Marine Ecosystems. *U.N. Atlas of the Oceans*; Garcia and Farmer (2000), LME #8: Scotian Shelf. *Large Marine Ecosystems of the World*; Halliday and Pinhorn (1990).

Page 94

Figure 6.4: ANGLER log, Captain Nathan Buck Jr., 1853. (NARA Waltham RG 36, Box 55, F448).

Page 95

affect the legendary abundance of species like cod: Vickers, D., editor (1995), *Marine Resources and Human Societies in the North Atlantic Since 1500*. ICER Conference Paper No. 5. Institute of Social and Economic Research, Memorial University, St. John's, Newfoundland.

the number of Beverly vessels fishing on the Scotian Shelf declined by half and the entire Beverly cod fleet by 55 percent: Rosenberg et al. (2005); Leavenworth (2006).

Page 96

CLARA M. PORTER: CLARA M. PORTER log, Captain Solomon Woodbury, 1853. (NARA Waltham RG36, Box 61: F462a).

price increased by about 25¢/quintal from the year before, and exports increased as well: p. 354–357 in O'Leary (1996).

Page 97

LODI: LODI log, Captain Samuel Wilson, 1857. (NARA Waltham RG36, Box 74: F496f).

ROBERT: ROBERT log, Captain R.C. Dennis, 1857. (NARA Waltham RG36, Box 84: F518f).

SUSAN CENTER: SUSAN CENTER log, Captain Thomas Gayton, 1857. (NARA Waltham RG36, Box 86: F526f & g).

precisely at the time that catches were declining: Leavenworth (2006).

Page 98

Figure 6.5: plates 23, 24, 26, 27, section V in Goode, G.B. (1884–1887), *The fisheries and fishery industries of the United States. Prepared through the co-operation of the commissioner of fisheries and the superintendent of the tenth census by George Brown Goode . . . and a staff of associates . . .* Government Printing Office, Washington, DC; Leavenworth (2006).

RICHMOND: RICHMOND log, Captain Abram A. Fisk, 1854. (NARA Waltham RG36, Box 82: F516d).

Page 99

Wilson expressed greater enthusiasm for boats: LODI log, Captain Samuel Wilson, 1856. (NARA Waltham R636, Box 74. F496e)

35 percent of Beverly vessels were using this new technology: Leavenworth (2006).

the well-financed and heavily subsidized French fleet of treaty violations by employing them: p. 29–30 in Sabine (1853).

French trawlers crossed the Laurentian Channel to Banquereau in 1858: Leavenworth (2006).

The effect was devastating on the already beleaguered Beverly fleet: Leavenworth (2006).

FRANKLIN: FRANKLIN log, Captain Enos Hatfield, 1858. (NARA Waltham RG36, Box 67: F474o).

LODI: LODI log, Captain Samuel Wilson, 1858. (NARA Waltham RG36, Box 74: F496g).

Page 100

PELICAN: PELICAN log, Captain Jones Hatfield, 1858. (NARA Waltham RG36, Box 80: F510g).

PRIZE BANNER: PRIZE BANNER log, Captain Solomon Woodbury, 1858. (NARA Waltham RG36, Box 82: F515b).

Tub trawling clearly made the fisheries more dangerous than ever for men on the banks: Leavenworth (2006).

the demand for fresh bait escalated as well: McKenzie, M.G. (2008), Baiting our memories: The impact of offshore technology change on inshore species around Cape Cod, 1860–1895. p. 77–89 in Starkey, D., P. Holm, and M. Barnard, editors *Oceans Past: Management Insights from the History of Marine Animal Populations*. London UK and Sterling, VA: *Earthscan*; Chapter 4 in McKenzie, M. (2011), *Clearing the Coastline, the 19th-Century Ecological and Cultural Transformation of Cape Cod*. Hanover, NH: University Press of New England.

PETREL: PETREL log, Captain Calvin Foster, 1854. (NARA Waltham RG36, Box 81: F512c).

Page 101

MAYFLOWER: MAYFLOWER log, Captain Gustavus Obear, 1858. (NARA Waltham RG36, Box 78: F503g).

EXCHANGE: EXCHANGE log, Captain Daniel McGrath, 1858. (NARA Waltham RG36, Box 65: F472h).

Wilson had experimented with them for almost a decade: for the technological shift from handlining to tub trawling, see p. 148–181, vol. 1, section V, in Goode, G.B. (1884–1887); ch. 12 in Innis (1940); Leavenworth (2006).

cod, and other demersal species would soon become "scarce as salmon": Bolster (2006).

Page 102

as modern fishers react when stocks become overfished: Holland, D.S., and J.G. Sutinen (2000), Location choice in New England trawl fisheries: Old habits die hard. *Land Economics* 76: 133–149.

"a few years since . . . [to] set line fishing (tub trawling), first practiced on it by the French and latterly by United States fishermen": p. 376–377 in Innis (1940).

"Not much fished at present by Americans": chart 4, vol. 3 in Goode (1884–1887).

greater opportunities enticed young men away from fishing: Collins, J.W. (1898), Decadence of the New England Deep Sea Fisheries. reprinted on p. 393–410 in Oppel, F., compiler, (1985), *Tales of the New England Coast*. Edison, NJ: Castle Books.

Page 103

the earliest regular scientific sampling surveys by thirty: Smith (1994).

based on total removals when cumulative effort is known: DeLury, D.B. (1947), On the estimation of biological populations. *Biometrics* 3: 145–167.

The Chapman-Delury method, developed in 1972: Chapman, D. (1974), Estimation of population size and sustainability yield of Sei whales in the Antarctic. *Report of the International Whaling Commission* 24: 82–90.

we estimated the biomass of the cod population in 1852, the first year of the time series: Rosenberg et al. (2005).

Page 104

Figure 6.6: Rosenberg et al. (2005).

Page 106

both large and small vessels fished inshore: Alexander, K.E., W.B. Leavenworth, J. Cournane, A.B. Cooper, S. Claesson, S. Brennan, G. Smith, L. Rains, K. Magness, R. Dunn, T.K. Law, R. Gee, W.J. Bolster, and A.A. Rosenberg (2009), Gulf of Maine cod in 1861: Historical analysis of fishery logbooks, with ecosystem implications. *Fish and Fisheries* 10: 428–449.

43 percent of all spoken vessels: Sources of data for the spoken vessels are as follows: NARA Waltham Record Group [RG] 36, Customs District Records: Barnstable MA District fishing agreements-Record #533; Marblehead MA District fishing agreements-Essex Institute Collection #9-11; Frenchman's Bay ME District logs and agreements Record #104, 105; Machias ME District logs-Record #76; fishing agreements-Record #75; Penobscot-Castine ME District logs Record #137; fishing agreements-Record #136. James Duncan Phillips Library, Peabody Essex Museum: Newburyport MA logs-Manuscript Collection [Mss] 282, Box 31; Barnstable MA District enrolled vessels; Marblehead MA District vessel registers; Applebee Collection of Castine ME vessels; Colcord, Lincoln, compiler, (1932),

List of vessels from Penobscot Bay, ME. in Wasson, George S. (1932), *Sailing Days on the Penobscot*. Salem, MA: Marine Research Society.

Page 107

Europeans had exploited those grounds since at least 1539: Ramusio 1565, quoted in Rosenberg et al. (2005).

much like the earliest stages of modern exploitation of an unfished species: Rosenberg et al. (2005).

adjusted for assumed rates of natural mortality (M) and recruitment: Rosenberg et al. (2005).

Page 108

estimated by robust regression techniques: Huber, P.J. (1981), *Robust Statistics*. New York: John Wiley & Sons.

the AUTODIFF language incorporated into AD Model Builder: Otter Research (1994), *AUTODIF: A C++ array language extension with automatic differentiation for use in nonlinear modeling and statistics*. Otter Research Ltd. Sidney, B.C., Canada. online at: http://otter-rsch.com/aut.zip; Otter Research (2004), *An introduction to AD Model Builder Version 5.0.1. for use in nonlinear modeling and statistics*. Otter Research Ltd. Sidney, B.C., Canada, online at: http://otter-rsch.com/adm.zip

Page 109

calculated in the 1950s by the Canadian government: Beatty, S.A., and H. Fougere (1957), *The Processing of Dried Salted Fish*. Ottawa: Fisheries Research Board of Canada; fn. 7 in Pope, P. (1995), Early Estimates: Assessments of catches in the Newfoundland cod fishery, 1660–1690. p. 9–40 in Vickers, D., editor *Marine Resources and Human Societies in the North Atlantic Since 1500*. St. Johns, Newfoundland: ICER.

French removals from 1858 and 1859: vol. 1 in Soublin, L. (1991), *Cent Ans de Pêche a Terre-Neuve*. Paris: Henri Veyrier.

carrying capacity of the Scotian Shelf for cod based on productivity data: Myers, R.A., B.R. MacKenzie, K.G. Bowen, and N.J. Barrowman (2001), What is the carrying capacity for fish in the ocean? A meta-analysis of population dynamics of North Atlantic cod. *Canadian Journal of Fisheries and Aquatic Science* 58: 1464–1476.

Page 110

the total carrying capacity of the Scotian Shelf 1.15 million metric tons: Myers et al. (2001).

the inshore banks near Nova Scotia, the Bay of Fundy, and waters as far south as Cape May, New Jersey: Mohn, R., L. Fanning, and W. MacEachern (1998), CSAS Res. Doc (98/78); DFO Canada 2000; Clark, D. (2002) personal communication (DFO, St. Andrews Biological Station); Fanning, L., R. Mohn, and W. MacEachern (2003), Canadian Science Advisory Secretariat [CSAS] Res. Doc.

(2003/027); Fanning, L. (2003), personal communication (DFO, Bedford Institute of Oceanography).

biomass for this area is less than 50,000 metric tons: Fanning, Mohn, and MacEachern (2003); Rosenberg et al. (2005).

Page 111

Figure 6.7: Rosenberg et al. (2005).

ecosystem-based management of fisheries resources: Pew Oceans Commission (2003), *America's Living Oceans: Charting a Course for Sea Change*, online at: http://www.pewtrusts.org/our_work_report_detail.aspx?id=30009&category=130; United States Commission on Oceans Policy (2004), *Preliminary Report of the U.S. Commission on Ocean Policy*, Governor's Draft. Washington, DC: U.S. Government Printing Office.

the trend in abundance we described for Scotian Shelf cod: Jackson et al. (2001); Pauly, D., and J.L. Maclean (2003), *In a Perfect Ocean: The State of Fisheries and Ecosystems in the North Atlantic Ocean*. Washington, DC: Island Press; Myers and Worm (2003); Hardt, M. (2008), Lessons from the past: The collapse of Jamaican coral reefs. *Fish and Fisheries* 10: 1–16.

Page 112

both history and biology indicate otherwise: Myers (2001); Northeast Fisheries Science Center (2002), *Final report of the working group on re-evaluation of biological reference points for New England groundfish*. Northeast Fisheries Science Center Reference Document 02-04; Rosenberg et al. (2005).

Chapter 7

Page 115

the social history of colonial New England's cod fishery: Vickers, D. (1994), *Farmers and Fishermen: Two Centuries of Work in Essex County, Massachusetts, 1630–1850*. Chapel Hill: University of North Carolina Press.

Page 116

rather than the question of its ecological sustainability: Ommer, R. (1991), *From Outpost to Outport: A Structural Analysis of the Jersey-Gaspé Cod Fishery, 1767–1886*. Montreal and Kingston: McGill-Queens University Press; Cadigan, S.T. (1995), *Hope and Deception in Conception Bay: Merchant-Settler Relations in Newfoundland, 1785–1855*. Toronto: University of Toronto Press.

undertook to measure some of these basic demographic phenomena: Laslett, P. (1963), Clayworth and Cogenhoe. p. 157–184 in Bell, H.E., and R.L. Ollard, editors *Historical Essays, 1600–1750, Presented to David Ogg*. New York: Barnes and Noble.

"we should have the chance of reconstructing the population of our country as it was during all those generations which went by before the census began in 1801, of doing it swiftly, accurately, and completely": p. 160 in Laslett (1963).

Page 117

Laslett discovered a number of astonishing facts: Laslett (1963).

households were a lot smaller (4.0–4.5 members) than Laslett had expected: Laslett (1963).

"We cannot yet tell: we may never be able to tell": p. 181–182 in Laslett (1963).

The World We Have Lost: Laslett, P. (1965), *The World We Have Lost*. New York: Scribner's.

Population History of England, 1541–1871: A Reconstruction. Wrigley, E.A., and R.S. Schofield (1981), *The Population History of England, 1541–1871: A Reconstruction*. London: Edward Arnold.

with its "more humane, much more natural relationships" was simply a myth: p. 236 in Laslett (1965).

Page 121

in this case the past is all we have to go on: Oreskes, N. (2000), Why predict? Historical perspectives on prediction in Earth Science. p. 23–40 in Sarewitz, D., R. Pielke, and R. Byerly, editors, *Prediction: Science, Decision Making, and the Future of Nature*. Washington, DC: Island Press.

marine ecology could well proceed along the same path: Safina, this volume.

drew from two mutually exclusive labor pools: Vickers (1994).

Page 122

how large were the cod stocks of that area in the age before the development of industrial fisheries?: Rosenberg, A.A, W.J. Bolster, K.E. Alexander, W.B. Leavenworth, A.B. Cooper, and M.G. McKenzie (2005), The history of ocean resources: Modeling cod biomass using historical records. *Frontiers in Ecology and the Environment* 3(2): 84–90; Leavenworth, W.B. (2006), Opening Pandora's Box: Tradition, competition and technology on the Scotian Shelf, 1852–1860. p. 29–49 in Starkey, D.J., and J.E Candow, editors *Studia Atlantica: Proceedings of the 7th Conference of the North Atlantic Fisheries History Association [NAFHA]* . NAFHA, Hull UK and Nordurslod IS; Bolster and Alexander, this volume.

to rebuild ecosystems as they were: Oreskes, N., and K. Belitz (2001), Philosophical Issues in Model Assessment. p. 23–41 in Anderson, M.G., and P.D. Bates, editors, *Model Validation: Perspectives in Hydrological Science*. Hoboken, NJ: John Wiley & Sons.

Page 123

our baseline definition of a pristine ecosystem seems to have fallen with each successive generation: Pauly, D. (1995), Anecdotes and the shifting baseline syndrome of fisheries. *Trends in Ecology and Evolution* 10: 430.

try to learn more about how they came into being than earlier generations ever have: Stille, A. (2002), *The Future of the Past*. New York: Farrar, Straus and Giroux.

Page 125

an example from New England's woods and rivers: Cronon, W. (1983), *Changes in the Land: Indians, Colonists, and the Ecology of New England*. New York: Hill and Wang.

Page 126

"if we leave them sufficient for their use we may lawfully take the rest, there being more than enough for them and us": Winthrop, J. (1629), Reasons to be Considered for . . . the Intended Plantation in New England. p. 70–74 in Heimert, A., and A. Delbanco, editors (1985), *The Puritans in America: A Narrative Anthology*. Cambridge, MA: Harvard University Press.

one needs to take much of what they wrote with a grain of salt: Cronon (1983).

alewives "pressing up in such shallow waters as will scarce permit them to swim": p. 65 in Wood, W. (1635), Vaughan, A.T., editor (1997), *New England's Prospect*. Amherst: University of Massachusetts Press.

shad "so thick . . . you could not put in your hand without touching some of them": Marston, P.M., and G. Myron (1938), Notes on Fish and Early Fishing in the Merrimack River System. p. 190 in Hoover, E.E., editor *Biological Survey of the Merrimack Watershed*. New Hampshire Fish and Game Department, Report #3.

mollusks were a lot smaller and much fewer in number than they had been at the time of European settlement: Kirby, M.X. (2004), Fishing down the coast: Historical expansion and collapse of oyster fisheries along continental margins. *Proceedings of the National Academy of Sciences of the United States of America* 101(35): 13096–13099.

transcends the peculiar biases or mistakes of individual sources: Jackson, J.B.C., M.X. Kirby, W.H. Berger, K.A. Bjorndal, L.W. Botsford, B.J. Bourque, R.H. Bradbury, R. Cooke, J. Erlandson, J.A. Estes, T.P. Hughes, S. Kidwell, C.B. Lange, H.S. Lenihan, J.M. Pandolfi, C.H. Peterson, R.S. Steneck, M.J. Tegner, and R.R. Warner (2001), Historical overfishing and the recent collapse of coastal ecosystems. *Science* 293: 629–637.

Page 127

"trophic interaction between species, habitat impacts of different gears, and a theory for dealing with the optimum placement and size of marine reserves": Pauly, D., V. Christensen, S. Guénette, T. Pitcher, U.R. Sumaila, C. Walters, R. Watson, and D. Zeller (2002), Toward sustainability in world fisheries. *Nature* 418: 689–695.

the fur trade in western Canada during the eighteenth century: Ray, A.J., and D. Freeman (1978), *'Give us Good Measure': An Economic Analysis of Relations*

Between the Indians and the Hudson's Bay Company Before 1763. Toronto: University of Toronto Press.

family-run mining operations in early modern England: Blanchard, I. (1978), Labour productivity and work psychology in the English mining industry, 1400–1600. *Economic History Review* 31: 1–24.

Page 128

economists and biologists have generated predictive theories of some power and considerable influence: Gordon, H.S. (1954), The economic theory of common property resource: The fishery. *Journal of Political Economy* 62: 124–142; Hardin, G. (1968), The tragedy of the commons. *Science* 162: 1243–1248.

what one can and cannot do with the resource: McCay, B.J. (1994–95), The ocean commons and community. *Dalhousie Review* 74: 310–338.

because their tribal culture was not inherently expansionist: Newell, D. (1995), Maritime Property Rights-An Historian's Perspective. p. 277–291 in Vickers, D., editor *Marine Resources and Human Societies in the North Atlantic Since 1500*. ISER Conference Paper Number 5. Institute of Social and Economic Research, St. John's, Newfoundland, Canada; Suttles, W. (1962), Variation in Habitat and Culture on the Northwest Coast. p. 26–44 in Suttles, W., editor (1987), *Coast Salish Essays*. Seattle: University of Washington Press; Suttles, W. (1968), Coping With Abundance: Subsistence on the Northwest Coast. p. 45–66 in Suttles (1987).

Page 129

"the march of folly," to borrow Barbara Tuchman's phrase: Tuchman, B.W. (1984), *The March of Folly: From Troy to Vietnam*. New York: Alfred A. Knopf; Sumaila and Pauly, this volume.

"the worst fisheries management failure in the world": p. 45 in Safina, C. (1997), *Song for the Blue Ocean*. New York: Henry Holt.

Page 130

subsidizing the fishing industry and the communities that pursued it: Sumaila and Pauly, this volume.

over which the island had no jurisdiction: Cadigan, S.T. (1995), "A 'Chilling Neglect': The British Empire and Colonial Policy on the Newfoundland Bank Fishery, 1815–1855." Paper presented to the Canadian Historical Association Annual Meeting, Montreal, August 1995.

Page 131

attract the imagination of Canada's ruling class or the majority of its voters: Ommer, R. (1985), What's wrong with Canadian fish? Journal of Canadian Studies 20: 122–142; Schrank, W.E., N. Roy, R. Ommer, and B. Skoda (1992), An inshore fishery: A commercially viable industry or an employer of last resort. *Ocean Development and International Law* 23: 335–367.

Part IV Introduction

Page 135

several recent review articles employ data from most of them: Jackson, J.B.C., M.X. Kirby, W.H. Berger, K.A. Bjorndal, L.W. Botsford, B.J. Bourque, R.H. Bradbury, R. Cooke, J. Erlandson, J.A. Estes, T.P. Hughes, S. Kidwell, C.B. Lange, H.S. Lenihan, J.M. Pandolfi, C.H. Peterson, R.S. Steneck, M.J. Tegner, and R.R. Warner (2001), Historical overfishing and the recent collapse of coastal ecosystems. *Science* 293: 629–637; Pandolfi, J.M., R.H. Bradbury, E. Sala, T.P. Hughes, K.A. Bjorndal, R.G. Cooke, D. McArdle, L. McClenachan, M.J.H. Newman, G. Paredes, R.R. Warner, and J.B.C. Jackson (2003), Global trajectories of the long-term decline of coral reef ecosystems. *Science* 301: 955–958; Lotze, H.K., H.S. Lenihan, B.J. Bourque, R. Bradbury, R.G. Cooke, M.C. Kay, S.M. Kidwell, M.X. Kirby, C.H. Peterson, and J.B.C. Jackson (2006), Depletion, degradation, and recovery potential of estuaries and coastal seas. *Science* 312: 1806–1809; Lotze, H.K., and B. Worm (2009), Historical baselines for large marine animals. *Trends in Ecology and Evolution* 24(5): 254–262.

Page 136

managers can better anticipate the likely environmental consequences of different actions: Lotze, H.K., and B. Worm (2009).

the carrying capacity of the environment and the biological characteristics of cod: Bolster and Alexander, this volume.

Chapter 8

Page 137

without historical baselines to use as reference points: Lotze, H.K., and B. Worm (2009), Historical baselines for large marine animals. *Trends in Ecology and Evolution* 24(5): 254–262.

our baseline for comparison depends on when we choose to measure it: Pauly, D. (1995), Anecdotes and the shifting baseline syndrome of fisheries. *Trends in Ecology and Evolution* 10: 430; Jackson, J.B.C., M.X. Kirby, W.H. Berger, K.A. Bjorndal, L.W. Botsford, B.J. Bourque, R.H Bradbury, R. Cooke, J. Erlandson, J.A. Estes, T.P. Hughes, S. Kidwell, C.B. Lange, H.S. Lenihan, J.M. Pandolfi, C.H. Peterson, R.S. Steneck, M.J. Tegner, and R.R. Warner (2001), Historical overfishing and the recent collapse of coastal ecosystems. *Science* 293: 629–637.

Page 138

changes in species and environments over centuries and millennia to millions of years: Committee on the Geologic Record of Biosphere Dynamics, and National Research Council (2005), *The Geological Record of Ecological Dynamics*. Washington, DC: The National Academies Press.

Page 139

Figure 8.1C: Ellis, R. (1991), *Men and Whales*. New York, Alfred A. Knopf.

Page 141

climate records based on stable isotopes or pollen: Committee on the Geologic Record of Biosphere Dynamics, and National Research Council (2005).

preserve bimonthly environmental records spanning millennia: Haug, G.H., D. Gunther, L.C. Peterson, D.M. Sigman, K.A. Hughen, and B. Aeschlimann (2003), Climate and the collapse of Maya civilization. *Science* 299: 1731–1735.

the oxygen isotope ^{18}O is a common proxy to determine past temperatures and reconstruct paleoclimates: Haug, G.H., and R. Tiedemann (1998), Effect of the formation of the Isthmus of Panama on Atlantic Ocean thermohaline circulation. *Nature* 393: 673–676; Zachos, J., M. Pagani, L. Sloan, E. Thomas, and K. Billups (2001), Trends, rhythms and aberrations in global climate 65 Ma to present. *Science* 292: 686–693.

river runoff over thousands of years: Ingram, B.L., and D. Sloan (1992), Strontium isotopic composition of estuarine sediments as paleosalinity-paleoclimate indicator. *Science* 255: 68–72; Haug et al. (2003).

seasonal climate cycles over intervals of several hundred years: Lough, J.M., and D.J. Barnes (1997), Several centuries of variation in skeletal extension, density and calcification in massive *Porites* colonies from the Great Barrier Reef: A proxy for seawater temperature and a background of variability against which to identify unnatural change. *Journal of Experimental Marine Biology and Ecology* 211: 29–67; Love, M.S., M.M. Yoklavich, B.A. Black, and A.H. Andrews (2007), Age of black coral (*Antipathes dendrochristos*) colonies, with notes on associated invertebrate species. *Bulletin of Marine Science* 80: 391–399; Roark, E.B., T.P. Guilderson, S. Flood-Page, R.B. Dunbar, B.L. Ingram, S.J. Fallon, and M. McCulloch (2005), Radiocarbon-based ages and growth rates of bamboo corals from the Gulf of Alaska. *Geophysical Research Letters* 32(L04606) (doi:10.1029/2004GL021919).

the environmental conditions when each layer was formed: Swart, P.K., and A. Grottoli (2003), Proxy indicators of climate in coral skeletons: A perspective. *Coral Reefs* 22(4): 313–315.

a 240-year record of precipitation and regional runoff in the Florida Keys: Swart, P.K., R.E. Dodge, and H.J. Hudson (1996), A 240-year stable oxygen and carbon isotopic record in a coral from South Florida: Implications for the prediction of precipitation in southern Florida. *Paalaios* 11(4): 362–375.

Figure 8.2: Norris, R.D. (1999), Hydrographic and tectonic control of plankton distribution and evolution. p. 173–194 in Abrantes, F., and A.C. Mix, editors *Reconstructing Ocean History: A window into the Future*. New York: Kluwer Academic/Plenum Publishers. Haug and Tiedemann (1998).

Page 142

being more abundant during glacial than interglacial periods: Norris, R.D. (1999).

multiple samples along environmental and geographic gradients: Roy, K., and J.M. Pandolfi (2005), Responses of marine species and ecosystems to past climate change. p. 160–175 in Lovejoy, T.E., and L. Hannah, editors *Climate Change and Biodiversity*. New Haven, CT: Yale University Press.

the population structure of reefs had remarkable stability over a period of ~100,000 years: Pandolfi, J.M. (1999), Response of Pleistocene coral reefs to environmental change over long time scales. *American Zoologist* 39: 113–130.

the widespread loss of the coral *Acropora* from the Caribbean in the 1990s had no parallel in the fossil record: Pandolfi, J.M., and J.B.C. Jackson (2006), Ecological persistence interrupted in Caribbean coral reefs. *Ecology Letters* 9: 818–826.

temporal distribution of coral rubble in sediment cores from Discovery Bay, Jamaica: Wapnick, C.M., W.F. Precht, and R.B. Aronson (2004), Millenial-scale dynamics of staghorn coral in Discovery Bay, Jamaica. *Ecology Letters* 7(4): 354–361.

Page 143

species dominance associated with the reduction of Colorado River flows in the early twentieth century: Committee on the Geologic Record of Biosphere Dynamics, and National Research Council (2005).

usually increased after permanent settlement and land clearing: McCulloch, M., S. Fallon, T. Wyndham, E. Hendy, J. Lough, and D. Barnes (2003), Coral record of increased sediment flux to the inner Great Barrier Reef since European settlement. *Nature* 421: 727–730.

water quality in Chesapeake Bay deteriorated quickly after permanent European settlement in the seventeenth and eighteenth centuries: Cooper, S.R., and G.S. Brush (1993), A 2,500-year history of anoxia and eutrophication in Chesapeake Bay. *Estuaries* 16: 617–626.

very good indicator of past anthropogenic change: Kidwell, S.M. (2007), Discordance between living and death assemblages as evidence for anthropogenic ecological change. *Proceedings of the National Academy of Sciences of the United States of America* 104: 17701–17706.

Figure 8.3: Cooper and Brush (1993).

Page 144

some groups have higher preservation probabilities than others: Valentine, J.W., D. Jablonski, S.M. Kidwell, and K. Roy (2006), Assessing the fidelity of the fossil record by using marine bivalves. *Proceedings of the National Academy of Sciences of the United States of America* 103: 6599–6604.

the environment of deposition and the extent of subsequent alteration: Kidwell, S.M., and K.W. Flessa (1995), The quality of the fossil record: Populations, species and communities. *Annual Review of Ecology and Systematics* 26: 269–299; Valentine et al. (2006).

estimating relative or absolute abundance compared to simple presence or absence: Kidwell and Flessa (1995).

this is sometimes possible under special circumstances: Chepstow-Lusty, A.,

J. Backman, and N.J. Shackleton (1989), Comparison of upper Pliocene Discoaster abundance variations from North Atlantic sites 552, 607, 658, 659, and 662: Further evidence for marine plankton responding to orbital forcing. *Proceedings of the Ocean Drilling Program, Scientific Results* 108: 121–141; Sexton, P., and R.D. Norris (2008), Dispersal and biogeography of marine plankton: Long distance dispersal of the foraminifer *Truncorotalia truncatulinoides*. *Geology* 36(11): 899–902.

Page 145

various strata indicates changes in their importance as nutritional resources over time: Erlandson, J.M., T.C. Rick, and R. Vellanoweth (2004), Human impacts on ancient environments: A case study from California's Northern Channel Islands. p. 51–83 in Fitzpatrick, S.M., editor *Voyages of Discovery: Examining the Past in Island Environments*. Westport CT: Praeger Press.

modern humans have heavily influenced coastal and island environments for much longer than previously believed: Rick, T.C., and J.M. Erlandson, editors (2008), *Human Impacts on Ancient Marine Ecosystems: a Global Perspective*. Berkeley and Los Angeles, University of California Press.

date back to the last interglacial, more than 100,000 years ago: Erlandson, J.M. (2001), The archaeology of aquatic adaptations: Paradigms for a new millennium. *Journal of Archaeological Research* 9: 287–350; Erlandson, J.M. (2002), Anatomically modern humans, maritime adaptations, and the peopling of the New World. p. 59–92 in Jablonski, N., editor *The First Americans: The Pleistocene Colonization of the New World*. San Francisco: California Academy of Sciences.

evidence for the colonization of the California coast by about 13,000 to 12,000 years ago: Erlandson (2001, 2002).

indicating resource depression by indigenous people: Broughton, J.M. (1997), Widening diet breadth, declining foraging efficiency, and prehistoric harvest pressure: Ichthyofaunal evidence from the Emeryville Shellmound, California. *Antiquity* 71: 845–862; Broughton, J.M. (2002), Prey spatial structure and behavior affect archaeological tests of optimal foraging models: Examples from the Emeryville Shellmound vertebrate fauna. *World Archaeology* 34: 60–83.

that indigenous people may have had little impact on this commonly used species: Steneck, R.S., and J.T. Carlton (2001), Human alterations of marine communities: Students beware! p. 445–468 in Bertness, M.D., S.D. Gaines, and M.E. Hay, editors *Marine Community Ecology*. Sunderland, MA: Sinauer Associates; Jackson et al. (2001).

increasing mesopredators (flounder, sculpin) between 4,350 and 400 years ago: Bourque, B.J., B.J. Johnson, and R.S. Steneck (2008), Possible prehistoric fishing effects on coastal marine food webs in the Gulf of Maine. p. 165–185 in Rick and Erlandson, editors (2008).

species that occurred in earlier times but are now regionally or globally extinct: Lotze, H.K., and I. Milewski (2004), Two centuries of multiple human impacts and successive changes in a North Atlantic food web. *Ecological Applications* 14(5): 1428–1447.

prior to the arrival of the Maori people in AD 1250: Smith, I. (2005), Retreat and resilience: Fur seals and human settlement in New Zealand. p. 6–18 in Monks, G., editor *The Exploitation and Cultural Importance of Sea Mammals*. Oxford UK: Oxbow Books.

Pages 146–147

Table 8.2: Rozaire, C. (1976), Archaeological investigations on San Miguel Island. unpublished manuscript in the Los Angeles County Museum of Natural History; Guthrie, D.A. (1980), Analysis of avifaunal and bat remains from midden sites on San Miguel Island. p. 689–702 in Power, D.M., editor *The California Islands: Proceedings of a Multidisciplinary Symposium*. Santa Barbara CA: Santa Barbara Museum of Natural History; Walker, P.L. (1980), Archaeological evidence for the recent extinction of three terrestrial mammals on San Miguel Island. p. 703–717 in Power, D.M., editor *The California Islands*; Rick, T.C., J.M Erlandson, and R.L. Vellanoweth (2001), Paleocoastal marine fishing on the Pacific Coast of the Americas: Perspectives from Daisy Cave, California. *American Antiquity* 66: 595–613.

their breeding range had shrunk to the southern part of the South Island: Smith (2005).

reduced range and numbers of fur seals until protection was implemented in 1873: Smith (2005).

Page 148

Figure 8.4: Erlandson, J.M., T.C. Rick, M.H. Graham, J.A. Estes, T.J. Braje, and R.L. Vellanoweth (2005), Sea otters, shellfish, and humans: 10,000 years of ecological interactions on San Miguel Island, California. p. 58-69 in Garcelon, D.K., and C.A. Schwemm, editors *Proceedings of the Sixth California Islands Conference*. Santa Barbara CA: Institute for Wildlife Studies and National Park Service.

collaborations of archaeologists and historical ecologists: Erlandson et al. (2005).

marine and terrestrial proteins leave different $^{13}C/^{12}C$ ratios: Richards, M.P., R.J. Schulting, and R.E.M. Hedges (2003), Sharp shift in diet at onset of Neolithic. *Nature* 425: 366.

Page 149

Figure 8.5: Broughton (1997, 2002).

except in areas of active tectonic uplift: Pandolfi, J.M. (1999), Response of Pleistocene coral reefs to environmental change over long time scales. *American Zoologist* 39: 113–130.

Page 150

Levels of neutral genetic variation increase with population size: Avise, J., R. Ball, and J. Arnold (1988), Current versus historical population sizes in vertebrate species with high gene flow: A comparison based on mitochondrial DNA lineages and inbreeding theory for neutral mutations. *Molecular Biology and Evolution* 5: 331–344.

populations on the order of 240,000, 360,000 and 265,000 individuals, respectively: Roman, J., and S.R. Palumbi (2003), Whales before whaling in the North Atlantic. *Science* 301: 508–510.

in evaluating the effects of whaling: Jackson, J., N. Patenaude, E. Carroll, and C.S. Baker (2008), How few whales were there after whaling? Inference from contemporary mtDNA diversity. *Molecular Ecology* 17: 236–251.

changes in the geographic distributions of species in response to past climatic change: e.g., Hewitt, G.M (2000), The genetic legacy of the Quaternary ice ages. *Nature* 405: 907–913; Hellberg, M.E., D.P. Balch, and K. Roy (2001), Climate-driven range expansion and morphological evolution in a marine gastropod. *Science* 292: 1707–1710; Wares, J.P., and C.W. Cunningham (2001), Phylogeography and historical ecology of the North Atlantic intertidal. *Evolution* 55: 2455–2469.

one of the oldest illustrations of whaling in existence: p. 91 in Ellis, R. (1991).

leading to modern fisheries and hunting statistics: Cushing, D.H. (1988), *The Provident Sea*. Cambridge UK, Cambridge University Press; Holm, P., T.D. Smith, and D.J. Starkey, editors (2001), *The Exploited Seas: New Directions for Marine Environmental History. Research in Maritime History* No. 21. St John's, Newfoundland, International Maritime Economic History Association/Census of Marine Life; Lotze, H.K. (2007), Rise and fall of fishing and marine resource use in the Wadden Sea, southern North Sea. *Fisheries Research* 87: 208–218.

Richard Hoffmann used a wide range of written and illustrated information to reconstruct fisheries in medieval Europe: Hoffmann, R.C. (1996), Economic development and aquatic ecosystems in medieval Europe. *The American Historical Review* 101: 631–669; Hoffmann, R.C. (2001), Frontier foods for late medieval consumers: Culture, economy, ecology. *Environment and History* 7: 131–167; Hoffmann, R.C. (2005), A brief history of aquatic resource use in medieval Europe. *Helgoland Marine Research* 59: 22–30.

Page 151

inventing aquaculture to raise carp, and expanding to saltwater fisheries: Hoffmann (1996, 2001, 2005).

Similar human responses have occurred many times and in many places to this day: Lotze, H.K. (2004), Repetitive history of resource depletion and management: the need for a shift in perspective. in Browman, H.I., and K.I. Stergiou, editors, Perspectives on ecosystem-based approaches to the management of marine resources. *Marine Ecology Progress Series* 274: 282–285.

seasonal distributions of whales in the ocean: Townsend, C.H. (1935), The distribution of certain whales as shown by logbook records of American whale ships. Zoologica 19: 1–50; Reeves, R.R., T.D. Smith, E. Josephson, P. Clapham, and G. Woolmer (2004), Historical observations of humpback and blue whales in the North Atlantic Ocean: Clues to migratory routes and possibly additional feeding grounds. *Marine Mammal Science* 20(4): 774–786.

very different from today's coastal migration routes to northern feeding grounds: Smith, T.D., J. Allen, P.J. Clapham, P.S. Hammond, S. Katona, F. Larsen, J. Lien, D. Mattila, P.J. Palsbøll, S. Sigurjónsson, P.T. Stevick, and N. Øien (1999), An ocean-basin-wide mark-recapture study of the North Atlantic humpback whale (*Megaptera novaeangliae*). *Marine Mammal Science* 15: 1–32.

the historical distribution of right whales in the North Pacific: Josephson, E.A., T.D. Smith, and R.R. Reeves (2008), Historical distribution of right whales in the North Pacific. *Fish and Fisheries* 9: 155–168.

the first signs of overfishing appeared in the mid-1800s: Hutchings, J.A., and R.A. Myers (1995), The biological collapse of Atlantic cod off Newfoundland and Labrador: An exploration of historical changes in exploitation, harvesting, technology and management. p. 38–93 in Arnason, R., and L. Felt, editors *The North Atlantic Fisheries: Successes, Failures and Challenges*. Charlottetown, Prince Edward Island, Canada: The Institute of Island Studies.

an increase in numbers of target species in the commercial fisheries: Lotze (2004, 2007); Lotze and Milewski (2004).

Page 152

Figure 8.6: Townsend (1935); Reeves et al. (2004); Smith et al. (1999).

extreme ecological degradation over the past century: Lotze, H.K. (2005), Radical changes in the Wadden Sea fauna and flora over the last 2,000 years. *Helgoland Marine Research* 59: 71–83.

loss of oyster banks and of complexity and diversity of benthic communities due to overfishing: Reise, K., E. Herre, and M. Sturm (1989), Historical changes in the benthos of the Wadden Sea around the island of Sylt in the North Sea. *Helgoland Marine Research* 43: 417–433.

Page 153

Figure 8.7: Lotze (2004); Lotze and Milewski (2004).

a common sign of eutrophication: Lotze and Milewski (2004).

changes in species composition began almost from the start of the trawl fishery a century before: Klaer, N.L. (2001), Steam trawl catches from south-eastern Australia from 1918 to 1957: trends in catch rates and species composition. *Marine and Freshwater Research* 52: 399–410.

fisheries data have become increasingly available since the late nineteenth to early twentieth centuries: Goode, G.B. (1884–1887), *The fisheries and fishery industries of the United States. Prepared through the co-operation of the commissioner of fisheries and the superintendent of the tenth census by George Brown Goode . . . and a staff of associates*. Washington, DC: Government Printing Office; Cushing (1988).

to infer past trends in abundance: Hilborn, R., and C.J. Walters (1992), *Quantitative Fisheries Stock Assessment: Choice, Dynamics and Uncertainty*. New York: Chapman and Hall; Smith, T.D. (1994), *Scaling Fisheries: The Science of Measuring the Effects of Fishing, 1855–1955*. Cambridge UK: Cambridge University Press.

Page 154

Sequential Population Analysis: Hilborn and Walters (1992).

Multi-Species Virtual Population Analysis: Hilborn and Walters (1992).

the trajectory of global sperm whale abundance from 1700 to 2000: Whitehead, H. (2002), Estimates of the current global population size and historical trajectory for sperm whales. *Marine Ecology Progress Series* 242: 295–304.

extinct Caribbean monk seals would have consumed a biomass of fish and invertebrates six times that found on typical reefs today: McClenachan, L., and A.B. Cooper (2008), Extinction rate, historical population structure and ecological role of the Caribbean monk seal. *Proceedings of the Royal Society* B; 275: 1351–1358.

population trends, spatial distribution, and productivity of fish stocks over time: Simon, J.E., and P.A. Comeau (1994), Summer distribution and abundance trends of species caught on the Scotian Shelf from 1970–1992, by research vessel groundfish survey. *Canadian Technical Report of Fisheries and Aquatic Sciences* 1953. Dartmouth NS, Department of Fisheries and Oceans; Smith, T.D. (2002), The Woods Hole bottom-trawl resource survey: Development of fisheries-independent multispecies monitoring. *ICES Marine Science Symposium* 215: 474–482.

how large predators affect community structure: Steneck and Carlton (2001); Worm, B., and R.A. Myers (2003), Meta-analysis of cod-shrimp interactions reveals top-down control in oceanic food webs. *Ecology* 84: 162–173.

the abundance and size of intertidal invertebrates: Griffiths, C.L., and G.M. Branch (1997), The exploitation of coastal invertebrates and seaweeds in South Africa: Historical trends, ecological impacts and implications for management. *Transactions of the Royal Society of South Africa* 52: 121–148, Castilla, J.C. (1999), Coastal marine communities: Trends and perspectives from human-exclusion experiments. *Trends in Ecology and Evolution* 14: 280–283.

Page 155

widespread declines in body sizes of rocky intertidal gastropods since the 1960s in southern California: Roy, K., A.G. Collins, B.J. Becker, E. Begovic, and J.M. Engle (2003), Anthropogenic impacts and historical decline in body size of rocky intertidal gastropods in southern California. *Ecology Letters* 6: 205–211.

changes in response to human harvesting have been documented for many other species: Fenberg, P.B., and K. Roy (2008), Ecological and evolutionary consequences of size-selective harvesting: How much do we know? *Molecular Ecology* 17: 209–220.

greatly altered by human disturbance before monitoring began: Baumgartner, T.R., A. Soutar, and V. Ferreira-Bartrina (1992), Reconstruction of the history of Pacific sardine and northern anchovy populations over the past two millennia from sediments of the Santa Barbara Basin, California. *California Cooperative Oceanic Fisheries Investigations Report* 33: 24–40; Jackson et al. (2001); Pandolfi, J.M., R.H. Bradbury, E. Sala, T.P. Hughes, K.A. Bjorndal, R.G. Cooke, D. McArdle, L. McClenachan, M.J.H. Newman, G. Paredes, R.R. Warner, and J.B.C. Jackson (2003),

Global trajectories of the long-term decline of coral reef ecosystems. Science 301: 955–958; Lotze and Milewski (2004).

the structure, diversity, and functioning of plant-dominated benthic communities: Worm, B., H.K. Lotze, H. Hillebrand, and U. Sommer (2002), Consumer versus resource control of species diversity and ecosystem functioning. *Nature* 417: 848–851.

"bottom up" and "top down" effects depend on each other and on the productivity of the system: Lotze, H.K., and B. Worm (2002), Complex interactions of climatic and ecological controls on macroalgal recruitment. *Limnology and Oceanography* 47: 1734–1741.

Page 156

Figure 8.8: Original specimens in museum collections are from the California Academy of Sciences, University of California Museum of Palaeontology, Los Angeles County Museum of Natural History, San Diego Natural History Museum, Santa Barbara Museum of Natural History and Scripps Institution of Oceanography Benthic Invertebrate Collection; recent field survey took place in Southern California from Los Angeles to San Diego. Roy et al. (2003).

due to overfishing during postwar fishing booms: Smith (1994).

Page 157

extinctions that occurred over the past two thousand years due to largescale human exploitation and habitat alteration: Wolff, W.J. (2000), The south-east North Sea: Losses of vertebrate fauna during the past 2000 years. *Biological Conservation* 95: 209–217.

severely reduced or extinct by 1900 due to human impacts: Lotze and Milewski (2004).

as has been demonstrated for waterbirds by using old maps of the Netherlands: van Eerden, M.R. (1997), *Patchwork: Patch use, habitat exploitation and carrying capacity for water birds in Dutch freshwater wetlands*. PhD thesis, Rijksuniversiteit Groningen, The Netherlands.

more than a hundred times greater than today: Lotze and Milewski (2004).

Page 158

despite great increases in effort and the efficiency and spatial extent of fisheries: Lotze and Milewski (2004).

sharks in the Gulf of Mexico have declined 90 to 99 percent between the 1950s and 1990s: Baum, J.K., and R.A. Myers (2004), Shifting baselines and the decline of pelagic sharks in the Gulf of Mexico. *Ecology Letters* 7: 135–145.

the same is true for large predatory fish communities from nine different study areas worldwide: Myers, R.A., and Worm, B. (2003), Rapid worldwide depletion of predatory fish communities. *Nature* 423: 280–283.

exploitation has fundamentally altered marine food webs and ecosystems over time: Jackson et al. (2001); Christensen, V., S. Guénette, J.J. Heymans, C.J. Walters, R. Watson, D. Zeller, and D. Pauly (2003), Hundred-year decline of North Atlantic predatory fishes. *Fish and Fisheries* 4: 1–24; Roman and Palumbi (2003); McClenachan and Cooper (2008).

the categorical abundance of different groups of reef organisms over different cultural periods: Pandolfi et al. (2003); Pandolfi, J.M., J.B.C. Jackson, N. Baron, R.H. Bradbury, H. Guzman, T.P. Hughes, F. Micheli, J. Ogden, H. Possingham, C.V. Kappel, and E. Sala (2005), Are US coral reefs on the slippery slope to slime? *Science* 307: 1725–1726.

patterns of decline have been documented for estuaries and coastal seas: Jackson et al. (2001); Lotze and Milewski (2004); Lotze, H.K., H.S. Lenihan, B.J. Bourque, R.H. Bradbury, R.G. Cooke, M.C. Kay, S.M. Kidwell, M.X. Kirby, C.H. Peterson, and J.B.C. Jackson (2006), Depletion, degradation, and recovery potential of estuaries and coastal seas. *Science* 312: 1806–1809.

powerful for estimating the decline in fish populations: Myers and Worm (2003).

relative contributions of fishing versus oceanographic change: McCall, this volume; Field et al., this volume.

understanding the effects of fishing on food webs: Worm, B., H.K. Lotze, and R.A. Myers (2004), Ecosystem effects of fishing and whaling in the North Pacific and Atlantic Ocean, p. 335–343 in Estes, J.A., R.L Brownell, D.P. DeMaster, D.F. Doak, and T.M. Williams, editors *Whales, Whaling and Ocean Ecosystems*. Berkeley and Los Angeles: University of California Press.

has now been done for fish stocks offshore in California and South Africa: Baumgartner et al. (1992); Baumgartner, T.R., U. Struck, and J. Alheit (2004), GLOBEC investigation of interdecadal to multi-centennial variability in marine fish populations. *PAGES News* 12: 19–21.

Page 159

Figure 8.9: Lotze and Milewski (2004).

correlations with fish stocks or other environmental variables can be established: Field, D.B., T.R. Baumgartner, C.D. Charles, V. Ferreira-Bartrina, and M.D. Ohman (2006), Planktonic foraminifera of the California Current reflect 20th-century warming. *Science* 311: 63–66.

better studied species whose diets are inferred to be similar: Pauly, D., V. Christensen, and C. Walters (2000), Ecopath, Ecosim, and Ecospace as tools for evaluating ecosystem impacts of fisheries. *ICES Journal of Marine Science* 57: 697–706; Jackson et al. (2001); Steneck and Carlton (2001); Lotze and Milewski (2004); Bascompte, J., C.J. Melián, and E. Sala (2005), Interaction strength combinations and the overfishing of a marine food web. *Proceedings of the National Academy of Sciences of the United States of America* 102: 5443–5447; McClenachan and Cooper (2008).

Page 160

Figure 8.10: Lotze et al. (2006); Lotze and Milewski (2004).

the phenomenon of "fishing down the food web" as well as overall dramatic losses in biomass: Pauly, D., V. Christensen, J. Dalsgaard, R. Froese, and F.C. Torres Jr. (1998), Fishing down marine food webs. *Science* 279: 860–863; Christensen et al. (2003); Pauly, D., and J.L. Maclean (2003), *In a Perfect Ocean: The State of Fisheries and Ecosystems in the North Atlantic Ocean*. Washington, DC: Island Press.

Chapter 9

Page 163

assures Fernand Braudel in his extensive economic history of Europe: Braudel, F. (1978), *The Structures of Everyday Life: The Limits of the Possible*. New York: Harper and Row.

evidence for economic stability in late sixteenth-century Venice: ch. 2 in Braudel (1978).

Page 164

Figure 9.1: ch. 2 in Braudel (1978).

Page 165

interpreting historical data about state of the oceans in the past requires making assumptions about the past: Vickers, this volume.

known as BALEEN II: Punt, A.E. (1999), A full description of the standard BALEEN II model and some variants thereof. *Journal of Cetacean Research and Management* 1(Suppl.): S267–S276.

Page 167

bracketed the number of sperm whales before whaling: Whitehead, H. (2002), Estimates of the current global population size and historical trajectory for sperm whales. *Marine Ecology Progress Series* 242: 295–304.

trajectories do not pass through well-known census values from recent decades: International Whaling Commission (2003), Report of the sub-committee on the comprehensive assessment of humpback whales. *Journal of Cetacean Research and Management* 5: 293–323.

known population parameters and the recorded hunting record: Henderson, D.A. (1984), Nineteenth-century gray whaling: Grounds, catches and kills, practices and depletion of the whale population. p. 159–186 in Jones, M.L., S.L. Swartz, and S. Leatherwood, editors *The Gray Whale* Eschrichtius robustus. Orlando FL, Academic Press.

do not combine to give a good fit of gray whale population trajectories to population counts in the twentieth century: Reilly, S. (1981), *Population Assessment and Population Dynamics of the California Gray Whale* (Eschrichtius robustus). Seattle: College of Fisheries, University of Washington.

complex models with changing carrying capacities are invoked: Punt, A.,

C. Allison, and G. Fay (2004), An examination of assessment models for the eastern North Pacific gray whale based on inertial dynamics. *Journal of Cetacean Research and Management* 6: 121–132.

bring historical and current population data into alignment: e.g., Punt et al. (2004).

loss rate of L=50 percent (1/(1-L) = 2.0) in preindustrial whaling in the North Atlantic: Mitchell, E., and R. Reeves (1983), Catch history, abundance, and present status of northwest Atlantic humpback whales. *Report International Whaling Commission* (special issue) 5: 153–212.

Peter B. Best and colleagues suggested L=25 percent (1/(1-L) = 1.3) for right whales: Best, P., J. Bannister, R. Brownell, and G.P. Donovan, editors (2001), *Right Whales: Worldwide Status*. London: International Whaling Commission.

J. E. Scarff suggested 57 percent (1/(1-L) = 2.3): Scarff, J.E. (2001), Preliminary estimates of whaling induced mortality in the 19th century North Pacific right whale (*Eubalaena japonica*) fishery adjusting for struck-but-lost whales and non-American whaling. p. 261–268 in Best et al., editors (2001).

the fate of the many whales that were chased and lost was seldom recorded: Sherman, S. (1984), The nature, possibilities and limitations of whaling logbook data. *Report International Whaling Commission* (special issue) 5: 35–39.

Page 168

Reeves noted that the quality of logbook entries often declined from the 1840s to the 1880s: Reeves, R. (1984), A note on interpreting historic log books and journals. *Report International Whaling Commission* (special issue) 5: 29.

recorded humpback whale catch in the North Atlantic of about 29,000 animals: Stevick, P., J. Allen, P. Clapham, N. Friday, S. Katona, F. Larsen, J. Lien, D. Mattila, P. Palsbøll, J. Sigurjónsson, T. Smith, N. Øien, and P. Hammond (2003), North Atlantic humpback whale abundance and rate of increase four decades after protection from whaling. *Marine Ecology Progress Series*258: 263–273.

"interpretation of the extracted information by scholars is essential": Sherman (1984).

Page 169

illuminate the past population sizes of hunted whale populations: Roman, J., and S.R. Palumbi (2003), Whales before whaling in the North Atlantic. *Science* 301: 508–510; Alter, S.E., E. Rynes, and S.R. Palumbi (2007), DNA evidence for historic population size and past ecosystem impacts of gray whales. *Proceedings of the National Academy of Sciences of the United States of America* 104(38): 15162–15167.

comparing DNA sequences that have been obtained from individuals in the same population: e.g., Baker, C., A. Perry, J. Bannister, M. Weinrich, R. Abernethy, J. Calambokidis, J. Lien, R. Lambertsen, J. Ramírez, O. Vasquez, P. Clapham, A. Alling, S. Obrien, and S. Palumbi (1993), Abundant mitochondrial DNA variation and worldwide population structure in humpback whales. *Proceedings of the National Academy of Sciences of the United States of America* 90(17): 8239–8243;

Baker, C., R. Slade, J. Bannister, R. Abernethy, M. Weinrich, J. Lien, J. Urban, P. Corkeron, J. Calmabokidis, O. Vasquez, and S. Palumbi (1994), Hierarchical structure of mitochondrial-DNA gene flow among humpback whales *megaptera novaeangliae*, worldwide. *Molecular Ecology* 3: 313–327.

where N_e is the long-term effective size of the population: Avise, J., R. Ball, and J. Arnold (1988), Current versus historical population sizes in vertebrate species with high gene flow: A comparison based on mitochondrial DNA lineages and inbreeding theory for neutral mutations. *Molecular Biology and Evolution* 5: 331–344.

current levels of whale diversity reflect past population sizes more than they reflect current levels: Baker et al. (1993).

Page 170

or 20–40 percent per million generations, in this region of the whale genome: Alter et al. (2007).

breeding females are thought to make up about one-sixth to one-eighth of a whale population: Roman and Palumbi (2003).

typically assumed in models and management plans: Alter et al. (2007).

the ratio of breeding to nonbreeding adult females are particularly critical: Roman and Palumbi (2003); Baker, C.S., and P.J. Clapham (2004), Modeling the past and future of whales and whaling. *Trends in Ecology and Evolution* 19: 365–371; Alter et al. (2007).

variation in reproductive success from adult to adult: see examples in Clutton-Brock, T.H., editor (1988), *Reproductive Success*. Chicago IL, Chicago University Press.

much smaller than the 50 percent assumed here: Nunney, L. (1993), The influence of mating system and overlapping generations on effective population size. *Evolution* 47:1329–1341; Nunney, L. (2000), The limits to knowledge in conservation genetics-the value of effective population size. *Evolutionary Biology* 32: 179–194.

instead of the 20–25 years observed today: Roman and Palumbi (2003).

Page 171

previous estimates of the mutation rate of one gene were low by about a factor of two: Alter, S.E., and S.R. Palumbi (2009), Comparing evolutionary patterns and variability in the mitochondrial control region and cytochrome b in three species of baleen whales. *Journal of Molecular Evolution* 68: 97–111.

genetic diversity in one population of North Pacific gray whale suggests a historic population of 90,000 instead of 20,000: Alter et al. (2007).

simulated the diversity of DNA sequences in gray whale populations as if they had been subjected to a bottleneck during whaling: Alter et al. (2007).

Page 172

together with the genetic data on humpback, fin, minke, and gray whales: Roman and Palumbi (2003); Alter et al. (2007); Ruegg, K.C., E.C. Anderson, C.S. Baker, M. Vant, J.A. Jackson, and S.R. Palumbi (2010), Are Antarctic minke whales unusually abundant because of 20th century whaling? *Molecular Ecology* 19: 281–291.

Figure 9.3: Alter et al. (2007).

Chapter 10

Page 178

some level of exploitation is sustainable for many biological populations: Rosenberg, A.A., M.J. Fogarty, M.P. Sissenwine, J.R. Beddington, and J.G. Shepherd (1993), Achieving sustainable use of renewable resources. *Science* 262: 828–829.

complex effects occur when fishing changes age structure, habitat, or subpopulation structure: Berkeley, S.A., M.A. Hixon, R.J. Larson, and M.S. Love (2004), Fisheries and sustainability via protection of age structure and spatial distribution of fish populations. *Fisheries* 29: 23–32; Robichaud, D., and G.A. Rose (2004), Migratory behavior and range in Atlantic cod: Inference from a century of tagging. *Fish and Fisheries* 5: 185–214.

widespread declines of fisheries resources worldwide: Jackson, J.B.C., M.X. Kirby, W.H. Berger, K.A. Bjorndal, L.W. Botsford, B.J. Bourque, R.H. Bradbury, R. Cooke, J. Erlandson, J.A. Estes, T.P. Hughes, S. Kidwell, C.B. Lange, H.S. Lenihan, J.M. Pandolfi, C.H. Peterson, R.S. Steneck, M.J. Tegner, and R.R. Warner (2001), Historical overfishing and the recent collapse of coastal ecosystems. *Science* 293: 629–637; Myers, R.A., and Worm, B. (2003), Rapid worldwide depletion of predatory fish communities. *Nature* 423: 280–283; Pauly, D., and J.L. Maclean (2003), *In a Perfect Ocean: The State of Fisheries and Ecosystems in the North Atlantic Ocean*. Washington, DC: Island Press; Ferretti, F., R.A. Myers, F. Serena, and H.K. Lotze (2008), Loss of large predatory sharks from the Mediterranean Sea. *Conservation Biology* 22(4): 952–964; Limburg, K.E., and J.R. Waldman (2009), Dramatic declines in North Atlantic diadromous fishes. *Bioscience* 59(11): 955–965; Worm, B., R. Hilborn, J.K. Baum, T.A. Branch, J.S. Collie, C. Costello, M.J. Fogarty, E.A. Fulton, J.A. Hutchings, S. Jennings, O.P. Jensen, H.K. Lotze, P.M. Mace, T.R. McClenahan, C. Minto, S.P. Palumbi, A.M. Parma, D. Richard, A.A. Rosenberg, R. Watson, and D. Zeller (2009), Rebuilding global fisheries. *Science* 325: 578–585.

Cod have been a mainstay of fisheries for much of the developed world for more than five hundred years: Kurlansky, M. (1997), *Cod: A Biography of the Fish That Changed the World*. New York: Walker and Co.

Page 179

with apparent trophic cascades affecting the whole ecosystem: Steneck, R.S. (1998), Human influences on coastal ecosystems: Does overfishing create trophic cascades? *Trends in Ecology and Evolution* 13: 429–430.

as recently shown by Rose for northern cod stocks: Rose, G.A. (2004), Reconciling overfishing and climate change with stock dynamics of Atlantic cod (*Gadus morhua*) over 500 years. *Canadian Journal of Fisheries and Aquatic Science* 61: 1553–1557.

what is the eventual "rebuilt" state of a system like the Gulf of Maine?: Rosenberg, A.A, W.J. Bolster, K.E. Alexander, W.B. Leavenworth, A.B. Cooper, and M.G. McKenzie (2005), The history of ocean resources: Modeling cod biomass using historical records. *Frontiers in Ecology and the Environment* 3(2): 84–90.

the migration of labor out of the community: McCay, B.J. (2001), Community based and cooperative solutions to the "fishermen's problem" in the Americas. p.175–194 in Burger, J., E. Ostrom, R.B. Norgaard, D. Policansky, and B.D. Goldstein, editors *Protecting the Commons: A Framework for Resource Management in the Americas*. Washington, DC: Island Press.

overfishing has been nearly unaffected or even exacerbated by management: Rosenberg, A.A. (2003), Managing to the margins: The overexploitation of fisheries. *Frontiers in Ecology and the Environment* 1: 102–106.

Page 180

unlikely to result in real recovery for "decades or centuries": MacCall, A.D. (2002), Fisheries-management and stock-rebuilding prospects under conditions of low-frequency environmental variability and species interactions. *Bulletin of Marine Science* 70(2): 613–628.

these resources are the "common heritage of mankind": United Nations Convention on the Law of the Sea (UNCLOS) (1982), Preamble. Convention. online at: http://www.un.org/Depts/los/convention_agreements/texts/unclos/closindx.htm

Page 181

slow response to the problem of overfishing and the loss of too many resources: Rosenberg (2003).

Page 182

a 95 percent decline in some resources: Myers and Worm (2003); Jennings, S., and J.L. Blanchard (2004), Fish abundance with no fishing: Predictions based on macroecological theory. *Journal of Animal Ecology* 73: 632–642; Rosenberg et al. (2005).

as major alteration of food webs: Jackson et al. (2001).

services like carbon sequestration: Link, J. (2002), Ecological considerations in fisheries management: When does it matter? *Fisheries* 27: 10–17.

recognize the need to conserve ecosystems, and not just target species: United Nations (1994), *The precautionary approach to fisheries with reference to the straddling fish stocks and highly migratory fish stocks*. United Nations, General Assembly, United Nations Conference on Straddling Fish Stocks and Highly Migratory Fish Stocks, New York, 14–31 March 1994. A/CONF. 164/INF/8 26 January 1994.

Every Atlantic cod stock has been overfished and depleted: Myers, R.A., and Worm, B. (2005), Extinction, survival or recovery of large predatory fishes. *Philosophical Transactions of the Royal Society* B 360: 13–20.

Page 183

fisheries management is undergoing a period of introspection: p. ix in Longhurst (2010), *Mismanagement of Marine Fisheries*. Cambridge and New York: Cambridge University Press.

developing aquaculture, especially for carp, and by regulating fishing gear along the coast: Hoffmann, R.C. (1995), Environmental change and the culture of common carp in medieval Europe. *Guelph Ichthyological Review* 3: 57–85; Hoffmann, R.C. (1996), Economic development and aquatic ecosystems in medieval Europe. *The American Historical Review* 101: 631–669; Hoffmann, R.C. (2001), Frontier foods for late medieval consumers: Culture, economy, ecology. *Environment and History* 7: 131–167; Hoffmann, R.C. (2005), A brief history of aquatic resource use in medieval Europe. *Helgoland Marine Research* 59: 22–30.

royal edicts regulated gear and imposed closures along the French coasts by the fourteenth century to stabilize valuable fish stocks: p. 24–25 in Roberts, C. (2007), *The Unnatural History of the Sea*. Washington, DC: Island Press.

Plimouth Plantation gave itself the legal authority to regulate fisheries: Anon. (1887), *Laws Relating to Inland Fisheries in Massachusetts 1623–1886*. Boston: Wright and Potter Printing Company, State Printers.

licensing privileges to catch and preserve certain fish: Leavenworth, W.B. (2008) The changing landscape of maritime resources in seventeenth-century New England. *International Journal of Maritime History* 20(1): 33–62.

Page 184

limited days fishing, established fishing seasons, protected spawning, issued licenses, imposed gear restrictions, and even instituted a landings moratorium—all before 1700: *Laws Relating to Inland Fisheries in Massachusetts 1623–1886* (1887).

a shortage of cod near Boston in the 1650s: Leavenworth (2008).

laissez faire policies favoring industrial fishing and factories won the day and undermined small producers: Judd, R.W. (1997), Common Lands, Common People: *The Origins of Conservation in Northern New England*. Cambridge, MA: Harvard University Press.

adapted to other species after the fact: Halliday, R.G., and A.T. Pinhorn (1990), The delimitation of fishing areas in the Northwest Atlantic. *Journal of Northwest Fishery Science* 10: 1–51.

162 million pounds of Atlantic herring were caught in coastal nets along the New England shore: p. 247–325 in Bowers, G.M. (1905), *Report of the Bureau of Fisheries, 1904*. Washington, DC: Government Printing Office.

extended past the continental shelf and covered half a million square kilometers: National Oceanic and Atmospheric Administration (2007), *Annual Commercial Landings Statistics*, online at: http://www.st.nmfs.noaa.gov/st1/commercial/landings/annual_landings.html; National Oceanic and Atmospheric Administration (2009), *Status of Fisheries Resources in the Northeastern US: Atlantic Herring*, online at: http://www.nefsc.noaa.gov/sos/spsyn/pp/herring/

developed from a substantial body of work by a wide range of scientists: p. 1–15 in Longhurst (2010).

Page 185

a policy decision to expedite fisheries treaties in the wake of World War II: Finley, C. (2009), The social construction of fishing, 1949. *Ecology and Society* 14(1): 6, online at: http://www.ecologyandsociety.org/vol14/iss1/art6/

the appeal of its "relative simplicity" and "lower information demand": p. 5 in Longhurst (2010).

They went from a parametric approach, which yielded an MSY of 16,600 metric tons, to a nonparametric approach, which yielded an MSY of 10,014 metric tons: Mayo, R., G. Sheperd, L. O'Brien, L. Col, and M. Traver (2008), *Gulf of Maine Cod. Groundfish Assessment Review Meeting III*. Woods Hole, MA: Northeastern Fisheries Science Center, online at: http://www.nefsc.noaa.gov/publications/crd/crd0815/pdfs/garm3f.pdf

Results indicated that monkfish were not overfished and overfishing was not occurring, although fish size had declined: Northeast Data Poor Stocks Working Group (2007), *Monkfish Assessment Summary for 2007*. Northeast Fisheries Science Center Reference Document 07-13, online at: http://www.nefsc.noaa.gov/publications/crd/crd0713/

"The assessment results released in August 2007 suggest[ed] that this [would] not be necessary": Haring, P., and J.-J. McGuire (2007), The monkfish fishery and its management in the northeastern USA. *ICES Journal of Marine Science* 65: 1370–1379.

biological indicators have been adjusted downward: Hermsen, J. (2009), Monkfish northern and southern fishery management area daily landings and days-at-sea limit allocations for FY 2011–FY 2013. *Draft Amendment 5 to the Monkfish FMP incorporating Stock Assessment and Fishery Evaluation as well as a DSEIS*. New England Fisheries Management Council, online at: http://www.nefmc.org/monk/index.html

some propose that the lobster fishery be its role model: Ames, T. (2010), Multispecies coastal shelf recovery plan: A collaborative, ecosystem-based approach. *Marine and Coastal Fisheries: Dynamics, Management, and Ecosystem Science* 2: 217–231.

catch has been relatively consistent: National Oceanic and Atmospheric Administration (2007), *Status of Fisheries Resources in the Northeastern US: American lobster*, online at: http://www.nefsc.noaa.gov/sos/spsyn/iv/lobster/

In 1887 Maine lobstermen averaged 200 lbs/trap: p. 286–323 in the U.S. Commission of Fish and Fisheries (1892), *Report of the Commissioner for 1888*. Washington, DC: US Government Printing Office.

in 2005 not quite 20 lbs/trap over the same grounds: Atlantic States Marine Fisheries Commission (2009), *American Lobster Stock Assessment Report for Peer Review*. Stock Assessment Report No. 09-01 (Supplement), online at: http://

www.asmfc.org/speciesDocuments/lobster/annualreports/stockassmtreports/2009 LobsterStockAssessmentReport.pdf

Page 186

Regulators are considering a five-year fishing moratorium to rebuild stocks: Lindsey, J. (2010), Regulators weigh 5-year southern NE lobstering ban. Associated Press, online via *News8* wtnh.com, at: http://www.wtnh.com/dpp/news /business/regulators-weigh-5-year-southern-ne-lobstering-ban

recovery from overfishing under adverse climate conditions may take a century: MacCall (2002).

there is no historical evidence that sustainable fisheries have ever been maintained for long periods of time: Longhurst (2010).

Page 187

an attempt to restore depleted groundfish resources: Murawski, S.A., R.W. Brown, S.X. Cadrin, R.K. Mayo, L. O'Brien, W.J. Overholz, and K.A. Sosebee (1999), New England Groundfish. p. 71–80 in *Our Living Oceans: Report on the Status of U.S. Living Marine Resources*, NOAA Technical Memorandum NMFSF/ SPO-41. Washington, DC: US Government Printing Office.

especially for cod, haddock, and yellowtail flounder: Murawski, S.A., R. Brown, H.-L. Lai, P.J. Rago, and L. Hendrickson (2000), Large-scale closed areas as a fishery management tool in temperate marine systems: The Georges Bank experience. *Bulletin of Marine Science* 66: 775–798; Murawski, S.A., P. Rago, and M. Fogarty (2004), Spillover effects from temperate marine protected areas. *American Fisheries Society Symposium* 42: 167–184.

closures helped to successfully rebuild Atlantic sea scallops and haddock on Georges Bank, although cod and yellowtail flounder are still recovering: Stokesbury, K.D.E. (2002), Estimation of sea scallop abundance in closed areas of Georges Bank, USA. *Transactions of the American Fisheries Society* 131(6): 1081–1092; Murawski et al. (2004); Rosenberg, A.A., J.H. Swasey, and M. Bowman (2006), Rebuilding US fisheries: Progress and problems. *Frontiers in Ecology and the Environment* 4(6): 303–308.

partly due to a very strong year-class in 2003: Brodziak, J., M.L. Traver, and L.A. Col (2008), The nascent recovery of the Georges Bank haddock stock. *Fisheries Research* 94: 123–132.

strong primary productivity: Friedland, K.D., J. Hare, G.B. Wood, L.A. Col, L.J. Buckley, D.G. Mountain, J. Kane, J. Brodziak, R.G. Lough, and C.H. Pilskaln (2008), Does the fall phytoplankton bloom control recruitment of Georges Bank haddock, *Melanogrammus aeglefinus*, through parental condition? *Canadian Journal of Fisheries and Aquatic Sciences* 65: 1076–1086.

density of the shellfish had increased to about half of the average harvest from 1977 to 1988: Stokesbury (2002).

biomass appears to be at levels not seen since the late 1960s: Stone, H.H., S. Gavaris, C.M. Legault, J.D. Neilson, and S.X. Cadrin (2004), Collapse and recovery of the yellowtail flounder (*Limanda ferruginea*) fishery on Georges Bank. *Journal of Sea Research* 51: 261–270.

important tools for ecosystem-based fisheries management: Pikitch, E.K, C. Santora, E.A. Babcock, A. Bakun, R. Bonfil, D.O. Conover, P. Dayton, P. Doukakis, D. Fluharty, B. Heneman, E.D. Houde, J. Link, P.A. Livingston, M. Mangel, M.K. McAllister, J. Pope, and K.J. Sainsbury (2004), Ecosystem-based fishery management. *Science* 305(5682): 346–347.

socioecological baselines required for successful ecosystem-based management: Halpern, B.S., S. Walbridge, K.A. Selkoe, C.V. Kappel, F. Micheli, C. D'Agrosa, J.F. Bruno, K.S. Casey, C. Ebert, H.E. Fox, R. Fujita, D. Heinemann, H.S. Lenihan, E.M.P. Madin, M.T. Perry, E.R. Selig, M. Spalding, R. Steneck, and R. Watson (2008), A global map of human impact on marine ecosystems. *Science* 319: 948–952.

Swept Area Seabed Impact (SASI) model: New England Fisheries Management Council (NEFMC) (2010), *The Swept Area Seabed Impact (SASI) Model*, online at: http://www.nefmc.org/habitat/council_mtg_docs/Jan%202010/100127_Habitat_SASI.pdf

Page 188

assessed the effect of bottom conditions on different fisheries, based on sampling surveys and fishermen's information: Goode, G.B. (1884–1887), *The fisheries and fishery industries of the United States. Prepared through the co-operation of the commissioner of fisheries and the superintendent of the tenth census by George Brown Goode . . . and a staff of associates*. Washington, DC: Government Printing Office; Rich, W.H. (1929), *Fishing Grounds of the Gulf of Maine*. Washington, DC: Government Printing Office.

Stellwagen Bank National Marine Sanctuary Management Plan of 2010 was informed by an assessment of historical marine ecology and socially constructed fishing tradition: MacDonald, C. (2010), *Stellwagen Bank National Marine Sanctuary Final Management Plan and Environmental Assessment*. Silver Spring, MD: NOAA Office of the National Marine Sanctuaries, online at: http://stellwagen.noaa.gov/management/fmp/pdfs/sbnms_fmp2010_lo.pdf

Resident and migrant mammals include five species of seals (harp, gray, harbor, hooded, and ringed seals), 10 species of whales (humpback, minke, fin, sperm, right, beluga, orca, sei, blue, and pilot), along with white-beaked, whitesided, common, bottlenose, and Risso's dolphins and harbor porpoise: p. 99–132 in MacDonald, C. (2010).

to under a million pounds in 2007: NOAA (2007), National Oceanic and Atmospheric Administration (2007) *Annual Commercial Landings Statistics. Website*, online at: http://www.st.nmfs.noaa.gov/st1/commercial/landings/annual_landings.html

Some scientists fear a total collapse is under way: Limburg and Waldman (2009).

Important to both freshwater and oceanic ecosystems: Garmen, G.C., and S.A.

Macko (1998), Contribution of marine-derived organic matter to an Atlantic coast, freshwater, tidal stream by anadromous clupeid fishes. *Journal of the North American Benthological Society* 17(3): 227–285; Saunders, R., M.A. Hachey, and C.W. Fay (2006), Maine's diadromous fish community: Past, present, and implications for Atlantic salmon recovery. *Fisheries* 31(11): 537–547.

Page 189

running through a gauntlet of man-made obstacles and dangers: Yako, L.A., M.E. Mather, and F. Juanes (2002), Mechanisms for migration of anadromous herring: An ecological basis for effective conservation. *Ecological Applications* 12(2): 521–534; Cieri, M., G. Nelson, and M.A. Armstrong (2008), Estimates of river herring bycatch in the directed Atlantic herring fishery. *Report prepared for the Atlantic States Marine Fisheries Commission*, Washington, DC., September 23, 2008; Limburg and Waldman (2009); Kritzer, J. (2009), Multi-scale ecology and multi-scale management of shad and river herring. *Ecosystem-Based Fishery Management in Chesapeake Bay, November 2009 Update*, Maryland Sea Grant, online at: http://www.mdsg.umd.edu/programs/policy/ebfm/update/nov09/

they return to the sea where they spend the majority of their lives: p. 118–125 in Colette, B.B., and G. Klein-MacPhee, editors (2002), *Bigelow and Schroeder's Fishes of the Gulf of Maine*, Third Edition. Washington, DC: Smithsonian Institution Press; Yako et al. (2002).

it helped reestablish a spawning population of two million alewives in just ten years: Natural Resources Council of Maine (2009), *Kennebec River Reborn 10 Years after Dam Removal*, online at: http://www.nrcm.org/newsdetail.asp?news=3128

a default closure of directed commercial fisheries in state waters by 2012 unless sustainable harvest plans have been approved: p. iii–vii in Atlantic States Marine Fisheries Commission Shad and River Herring Plan Development Team (2010), *Shad and River Herring Management Plan*, online at: http://www.asmfc.org/"managed species-shad and river herring"

significant river herring bycatch occurs in the herring and mackerel fisheries, among others: Shepherd, G. (1986), *Evaluation of the river herring by-catch in the mackerel fishery*. Woods Hole Laboratory Reference Document 86-10. U.S. Department of Commerce; Cieri et al. (2008); Wigley, S.E., J. Blaylock, and P.J. Rago (2009), *River herring discard estimation, precision and sample size analysis*. Northeast Fish Science Center Reference Document 09-20. U.S. Department of Commerce.

the magnitude of bycatch likely varies because of unique at-sea migration patterns of genetically distinct stocks: Colette and Klein-MacPhee (2002); Haas-Castro, R. (2006), *River Herring. Status of Fishery Resources off the Northeastern U.S.* NEFSC, U.S. Department of Commerce; Cournane, J.M., and S.J. Correia (2010), *UPDATE: Identification of river herring hotspots at sea using fisheries dependent and independent datasets, including Supplemental Material*. Report prepared for the New England Fisheries Management Council Plan Development Team.

locations and timing of river herring hot spots in New England waters have been modeled using recent data: Cournane and Correia (2010).

Page 190

given willingly by fishermen who were eager to work with the commission: p. 9 in Bigelow, H.B. and W.W. Welsh (1925), *Fishes of the Gulf of Maine*. Washington, DC: Government Printing Office.

concentrates along those boundaries may cause local depletion of fish stocks: Murawski, S.A., S. Wigley, M. Fogarty, P. Rago, and D. Mountain (2005), Effort distribution and catch patterns adjacent to temperate MPAs. *ICES Journal of Marine Science* 62(6): 1150–1167.

incentives for stewardship and limit perverse incentives to overharvest: Steneck, R.S., and J.A. Wilson (2010), A fisheries play in an ecosystem theater: Challenges of managing ecological and social drivers of marine fisheries at multiple spatial scales. *Bulletin of Marine Science* 86(2): 387–4120; Ames (2010).

Page 191

attempts to rationalize competing needs and interests while maximizing human well-being and ecosystem integrity: Lorenzen, K., R.S. Steneck, R.R. Warner, A.M. Parma, F.C. Coleman, and K.M. Leber (2010), The spatial dimensions of fisheries, putting it all in place. *Bulletin of Marine Science* 86(2): 169–177.

Chapter 11

Page 193

when expansion was no longer possible, collapse was even more sudden and severe: Tainter, J. (1988), *The Collapse of Complex Societies*. Cambridge UK: Cambridge University Press; Redman, C.L. (1999), *Human Impact on Ancient Environments*. Tucson: University of Arizona Press; Gill, R.B. (2000), *The Great Maya Droughts: Water, Life, and Death*. Albuquerque: University of New Mexico Press; Diamond, J. (2005), *Collapse: How Societies Choose to Fail or Succeed*. New York: Viking.

Collapse of marine ecosystems and fisheries has followed a strikingly similar pattern: Jackson, J.B.C., M.X. Kirby, W.H. Berger, K.A. Bjorndal, L.W. Botsford, B.J. Bourque, R.H. Bradbury, R. Cooke, J. Erlandson, J.A. Estes, T.P. Hughes, S. Kidwell, C.B. Lange, H.S. Lenihan, J.M. Pandolfi, C.H. Peterson, R.S. Steneck, M.J. Tegner, and R.R. Warner (2001), Historical overfishing and the recent collapse of coastal ecosystems. *Science* 293: 629–637; Pauly, D., V. Christensen, J. Dalsgaard, R. Froese, and F.C. Torres Jr. (1998), Fishing down marine food webs. *Science* 279: 860–863; Pauly, D., and J.L. Maclean (2003), *In a Perfect Ocean: The State of Fisheries and Ecosystems in the North Atlantic Ocean*. Washington, DC: Island Press; as well as Sumaila and Pauly in chapter 2, and Parts II and III, this volume.

as in the case of cod: Pauly and Maclean (2003); Part III, this volume.

fishers move on to other species, which are generally smaller and grow faster: Bonfil, R., G. Munro, U.R. Sumaila, H. Valtysson, M. Wright, T. Pitcher,

D. Preikshot, N. Haggan, and D. Pauly (1998), Distant water fleets: An ecological, economic and social assessment. *Fisheries Centre Research Reports* 6(6). Fisheries Centre, University of British Columbia, Canada; Pauly et al. (1998); Pauly, D., and M.-L. Palomares (2005), Fishing down marine food web: It is far more pervasive than we thought. *Bulletin of Marine Science* 76(2): 197–211.

overfishing is repeated over and over until there are almost no fish left: Peterson, G.D., S.R. Carpenter, and W.A. Brock (2003), Uncertainty and the management of multistate ecosystems: An apparently rational route to collapse. *Ecology* 84: 1403–1411.

Page 194

Atlantic bluefin tuna provide a fascinating example of this shifting baselines problem in fisheries: Pauly, D. (1995), Anecdotes and the shifting baseline syndrome of fisheries. *Trends in Ecology and Evolution* 10:430.

in the mid-1600s a single trap on the southern coast of the Iberian Peninsula caught up to 100,000 tuna every year: unpublished data, Medina-Sidonia family archives.

catches began to decrease several centuries ago: unpublished data, Medina-Sidonia family archives; Florido, D. (2004), The bio-economic crisis of "Almadraba" (tuna trap-net) fishing from the 14th to the 19th centuries: Economic, social, political and ideological factors. paper presented at the HMAP-Mediterranean workshop (Barcelona, 20–23 September 2004).

Page 195

Figure 11.2: Unpublished data, Medina-Sidonia family archives.

the data are more fragmentary and recent: Bolster, Alexander and Leavenworth, Lotze et al., and Palumbi, this volume.

long-term oscillations in sardines and anchovies driven primarily by oceanographic change: Part II, this volume.

International Commission for the Conservation of Atlantic Tunas (ICCAT) still assumes that the few remaining tuna can be fished to achieve a maximum sustainable yield: Ravier, C., and J.-M. Fromentin (2001), Long-term fluctuations in the eastern Atlantic and Mediterranean bluefin tuna population. *ICES Journal of Marine Science* 58: 1299–1317.

a measure that would have banned international trade in this endangered species: Stokstad, E. (2010), Trade trumps science for marine species at international meeting. *Science* 328: 26–27.

extreme degradation of ecosystems today: Jackson, J.B.C. (2008), Ecological extinction and evolution in the brave new ocean. *Proceedings of the National Academy of Sciences of the United States of America* 105(supplement 1): 11458–11465.

when the ocean's riches were viewed as inexhaustible: Huxley, T.H. (1884), Inaugural address of the fishery conferences. *Fisheries Exhibition Literature* 4: 1–19.

the waters all around New York City were a dead zone: Waldman, J. (2000), *Heartbeats in the Muck: A Dramatic Look at the History, Sea Life, and Environment of New York Harbor*. New York: The Lyons Press.

Page 196

limits our ability to do much about it: Jackson et al. (2001); Knowlton, N. (2001), The future of coral reefs. *Proceedings of the National Academy of Sciences of the United States of America* 98: 5419–5425; Pandolfi, J.M., R.H. Bradbury, E. Sala, T.P. Hughes, K.A. Bjorndal, R.G. Cooke, D. McArdle, L. McClenachan, M.J.H. Newman, G. Paredes, R.R. Warner, and J.B.C. Jackson (2003), Global trajectories of the long-term decline of coral reef ecosystems. *Science* 301: 955–958; Pandolfi, J.M., J.B.C. Jackson, N. Baron, R.H. Bradbury, H. Guzman, T.P. Hughes, C.V. Kappel, F. Micheli, J. Ogden, H. Possingham, and E. Sala. (2005), Are US coral reefs on the slippery slope to slime? *Science* 307: 1725–1726; Knowlton, N., and J.B.C. Jackson (2008), Shifting baselines, local impacts, and global change on coral reefs. *PloS Biology* 6: 215–220 (doi10.1371/journal.pbio.0060054).

for several thousand to hundreds of thousands of years before their rapid demise in the twentieth century: Pandolfi, J.M. (2002), Coral community dynamics at multiple scales. *Coral Reefs* 21: 13–23; Aronson, R.B., I.G. Macintyre, W.F. Precht, T.J.T. Murdoch, and C.M. Wapnick (2002), The expanding scale of species turnover events on coral reefs in Belize. *Ecological Monographs* 72: 233–249; Aronson, R.B., I.G. Macintyre, C.M. Wapnick, and M.W. O'Neill (2004), Phase shifts, alternative states, and the unprecedented convergence of two reef systems. *Ecology* 85: 1876–1891; Pandolfi, J.M., and J.B.C. Jackson (2006), Ecological persistence interrupted in Caribbean coral reefs. *Ecology Letters* 9: 818–826.

reefs were already mildly to severely degraded: Pandolfi et al. (2003).

with seaweeds commonly taking over from the corals: Hughes, T.P. (1994), Catastrophes, phase shifts and large-scale degradation of a Caribbean coral reef. *Science* 265: 1547–1551; Gardner, T.A., I.M. Côté, J.A. Gill, A. Grant, and A.R. Watkinson (2003), Long-term region-wide declines in Caribbean corals. *Science* 301: 958–960; Wilkinson, C.R., editor (2004), *Status of the reefs of the world: 2004*. Townsville, Australia: Global Reef Monitoring Network and Australian Institute of Marine Science; Bruno, J.F., and E.R. Selig (2007), Regional decline of coral cover in the Indo-Pacific: Timing, extent, and subregional comparisons. *PloS One* 2:e711 (doi:10.1371/journal.pone.0000711).

Abundance of reef fishes has declined by 90–95 percent just in the last fifty years: Gardner et al. (2003).

sea turtles by more than 99 percent: McClenachan, L., J.B.C. Jackson, and M.J.H. Newman (2006), Conservation implications of historic sea turtle nesting beach loss. *Frontiers in Ecology and Environment* 4: 290–296.

yield great differences in coral and fish populations: Friedlander, A.M., and E.E. DeMartini (2002), Contrasts in density, size, and biomass of reef fishes between the northwestern and the main Hawaiian Islands: The effects of fishing down apex predators. *Marine Ecology Progress Series*230: 253–264; Knowlton and Jackson (2008); Sandin, S.A., J.E. Smith, E.E. DeMartini, E.A. Dinsdale, S.D. Donner, A.M. Friedlander, T. Konotchick, M. Malay, J.E. Maragos, D. Obura, O. Pantos, G. Paulay, M. Richie, F. Rohwer, R.E. Schroeder, S. Walsh, J.B.C. Jackson, N. Knowlton, and E. Sala (2008), Baselines and degradation of coral reefs in the northern Line Islands. *PloS One* 3(2): e1548 (doi:10.1371/journal.pone.0001548).

all are officially listed as endangered species by the IUCN: Meylan, A., and M. Donnelly (1999), Status justification for listing the hawksbill turtle (*Eretmochelys imbricata*) as critically endangered on the 1996 IUCN Red List of threatened animals. *Chelonian Conservation and Biology* 3:200–224; Seminoff, J.A. (2002), *IUCN Red List Global Status Assessment: Green Turtle* (Chelonia mydas). Gland, Switzerland: IUCN Marine Turtle Specialist Group; Bjorndal, K.A., and J.B.C. Jackson (2003), Roles of sea turtles in marine ecosystems: Reconstructing the past. p. 259–273, vol. II in Lutz, P.L., J.A. Musick, and J. Wyneken, editors *The Biology of Sea Turtles*. Boca Raton, FL: CRC Press; International Union for Conservation of Nature and Natural Resources (IUCN) (2008), *IUCN Red List of Threatened Species*, online at: http://www.redlist.org; McClenachan et al. (2006).

on the verge of extinction in the Mediterranean and the Hawaiian Islands: Ragen, T.J., and D.M. Lavigne (1999), The Hawaiian monk seal: Biology of an endangered species. p. 224–245 in Twiss, J.R. Jr., and R.R. Reeves, editors *Conservation and Management of Marine Mammals*. Washington, DC: Smithsonian Institution Press; Pires, R., H. Costa Neves, and A.A. Karamanlidis (2008), The critically endangered Mediterranean monk seal *Monachus monachus* in the archipelago of Madeira: Priorities for conservation. *Oryx* 42: 278–285; McClenachan, L., and A.B. Cooper (2008), Extinction rate, historical population structure and ecological role of the Caribbean monk seal. *Proceedings of the Royal Society B* 275: 1351–1358 (doi:10.1098/rspb.2007.1757).

among myriad other factors: Hughes (1994); Knowlton, N. (2001); Jackson et al. (2001); Hughes, T.P., A.H. Baird, D.R. Bellwood, M.Card, S.R. Connolly, C. Folke, R. Grosberg, O. Hoegh-Guldberg, J.B.C. Jackson, J. Kleypas, J.M. Lough, P. Marshall, M. Nyström, S.R. Palumbi, J.M. Pandolfi, B. Rosen, and J. Roughgarden (2003), Climate change, human impacts, and the resilience of coral reefs. *Science* 301: 929–933; Pandolfi et al. (2003, 2005); Knowlton and Jackson (2008).

allows seaweeds to rapidly increase at the expense of corals: Hughes (1994); Bellwood, D.R., T.P. Hughes, and A.S. Hoey (2006), Sleeping functional group drives coral-reef recovery. *Current Biology* 16: 2434–2439.

by overgrowth or indirectly by promoting coral disease: Hughes (1994).

hence outbreaks of disease: Nugues, M.M., G.W. Smith, R.J. van Hooidonk, M.I. Seabra, and R.P.M. Bak (2004), Algal contact as a trigger for coral disease. *Ecology Letters* 7: 919–923; Kline, D.I., N.M. Kuntz, M. Breitbart, N. Knowlton, and F. Rohwer (2006), Role of elevated organic carbon levels and microbial activity in coral mortality. *Marine Ecology Progress Series*314: 119–125; Smith, J.E.,

M. Shaw, R.A. Edwards, D. Obura, O. Pantos, E. Sala, S.A. Sandin, S. Smirga, M. Hatay, and F. Rohwer (2006), Indirect effects of algae on coral: Algae-mediated, microbe-induced coral mortality. *Ecology Letters* 9: 835–845.

Page 197

mass mortality of corals: Hughes et al. (2003).

Coral reefs as we know them may simply disappear: Hoegh-Guldberg, O., P.J. Mumby, A.J. Hooten, R.S. Steneck, P. Greenfield, E. Gomez, C.D. Harvell, P.F. Sale, A.J. Edwards, K. Caldeira, N. Knowlton, C.M. Eakin, R. Iglesias-Prieto, N. Muthiga, R.H. Bradbury, A. Dubi, M.E. Hatziolos (2007), Coral reefs under rapid climate change and ocean acidification. *Science* 318: 1737–1742.

nearly complete in Barbados in the early twentieth century: Lewis, J.B. (1984), The *Acropora* inheritance: A reinterpretation of the environment of fringing reefs in Barbados, West Indies. *Coral Reefs* 3: 117–122.

show comparable declines at sites scattered throughout the Caribbean before the 1980s: Jackson et al. (2001).

bleaching increased to epidemic proportions: Hughes (1994); Aronson et al. (2002); Aronson et al. (2004).

but have little bearing upon the causes or magnitude of changes that may have occurred earlier elsewhere, as in the case of Barbados: Lewis (1984).

unless we reduce all the threats as soon as possible: Pandolfi et al. (2005); Jackson, J.B.C., J.C. Ogden, J.M. Pandolfi, N. Baron, R.H. Bradbury, H.M. Guzman, T.P. Hughes, C.V. Kappel, F. Micheli, H.P. Possingham, and E. Sala (2005), Reassessing U.S. coral reefs—response. *Science* 308: 1740–1742; Knowlton and Jackson (2008).

unless we also halted overfishing and the runoff of pollution from the land: Hughes et al. (2003); Sandin et al. (2008); Knowlton and Jackson (2008).

Page 198

perhaps to increase fisheries yields: Roberts, C.M., J.A. Bohnsack, F. Gell, J.P. Hawkins, and R. Goodridge (2001), Effects of marine reserves on adjacent fisheries. *Science* 294: 1920–1923; Lubchenco, J., S.R. Palumbi, S.D. Gaines, and S. Adelman (2003), Plugging a hole in the ocean: The emerging science of marine reserves. *Ecological Applications* 13(supplement): S3–S7; Norse, E.A., C.B. Grimes, S. Ralston, R. Hilborn, J.C. Castilla, S.R. Palumbi, D. Fraser, and P. Kareiva (2003), Marine reserves: The best option for our oceans? *Frontiers in Ecology and the Environment* 1: 495–502.

the monk seal was already extinct: Jackson (1997); Jackson et al. (2001); McClenachan et al. (2006); McClenachan and Cooper (2008).

destabilize reef food webs and community structure: Bascompte, J., C.J. Melián, and E. Sala (2005), Interaction strength combinations and the overfishing of a marine food web. *Proceedings of the National Academy of Sciences of the United States of America* 102: 5443–5447; Sandin et al. (2008).

wipes out entire meadows and their associated shrimp and fish stocks: Jackson et al. (2001).

water quality has declined severely: Pandolfi et al. (2005).

extreme overfishing of reef fishes to the point of ecological extinction should have sounded the alarm: Hughes (1994); Knowlton, N. (1992), Thresholds and multiple stable states in coral reef community dynamics. *American Zoologist* 32: 674–682; Knowlton, N. (2004), Multiple "stable" states and the conservation of marine ecosystems. *Progress in Oceanography* 60:387-396; Jackson (1997).

Page 199

overfishing had upset this balance by the early 1900s: Hardt, M.J. (2009), Lessons from the past: the collapse of Jamaican coral reefs. *Fish and Fisheries* 10: 143–158.

identify conservation priorities to ward off imminent collapse: Pandolfi et al. (2005).

incorporate the data into a historical database: sensu Pandolfi et al. (2003, 2005).

an entirely different trajectory to a new alternative community state: Knowlton 2004.

"You can't go home again": Wolfe, T. (1940), *You Can't Go Home Again*. New York: Harper and Row.

Page 200

Figure 11.3: Pandolfi et al. (2003); Pandolfi et al. (2005), Knowlton (2004).

Coral reef communities were remarkably stable for millennia until their geologically instantaneous collapse: Pandolfi (1996, 2002); Aronson et al. (2002, 2004); Pandolfi and Jackson (2007).

cyclical, and therefore predictable, fluctuations over comparable time periods: MacCall, this volume.

to distinguish signal from noise: Jackson, J.B.C., and K.G. Johnson (2001), Measuring past biodiversity. *Science* 293: 2401–2403.

seem particularly promising for analysis of pelagic baseline communities: Lotze et al., this volume.

Page 201

whether everything happened all at once when some critical threshold was reached: Scheffer, M., S. Carpenter, J.A. Foley, C. Folke, and B. Walker (2001), Catastrophic shifts in ecosystems. *Nature* 413: 591–596; Knowlton (2004).

they may have an oceanographic as well as anthropogenic explanation: Part 2, this volume.

human impacts must have been primarily responsible: Jackson et al. (2001); Pandolfi et al. (2003); Steadman, D.W., P.S. Martin, R.D.E MacPhee, A.J.T. Jull, H.G. McDonald, C.A. Woods, M. Iturralde-Vinent, and G.W.L. Hodgins (2005),

Asynchronous extinction of Late Quaternary sloths on continents and islands. *Proceedings of the National Academy of Sciences of the United States of America* 102: 11763–11768.

in the loss of resilience leading up to catastrophic shifts: Hughes (1994); Scheffer et al. (2001); Knowlton (2004); Hughes, T.P., M.J. Rodrigues, D.R. Bellwood, D. Ceccarelli, O. Hoegh-Guldberg, L. McCook, N. Moltschaniwskyj, M. S. Pratchett, R.S. Steneck, and B. Willis (2007), Phase shifts, herbivory, and the resilience of coral reefs to climate change. *Current Biology* 17: 360–365.

extreme natural disturbances such as especially strong hurricanes: Woodley, J.D., E.A. Chornesky, P.A. Clifford, J.B.C. Jackson, L.S. Kaufman, N. Knowlton, J.C. Lang, M.P. Pearson, J.W. Porter, M.C. Rooney, K.W. Rylaarsdam, V.J. Tunnicliffe, C.M. Wahle, J.L. Wulff, A.S.G. Curtis, M.D. Dallmeyer, B.P. Jupp, M.A.R. Koehl, J. Neigel, and E.M. Sides (1981), Hurricane Allen's impact on Jamaican coral reefs. *Science* 214: 749–755.

oceanographic regime shifts: MacCall, this volume.

push anthropogenically stressed ecosystems beyond the point of no return: Hsieh, C., S.M. Glaser, A.J. Lucas, and G. Sugihara (2005), Distinguishing random environmental fluctuations from ecological catastrophes for the North Pacific Ocean. *Nature* 435: 336–340; Jackson et al. (2005).

Page 202

the distribution of biomass at different trophic levels of the food web: Jackson, J.B.C., and E. Sala (2001), Unnatural oceans. *Scientia Marina* 65 (supplement): S273–S281; Jackson, J.B.C. (2006). When ecological pyramids were upside down. p. 27–37 in Estes, J.A., editor *Whales, Whaling, and Ocean Ecosystems*. Berkeley and Los Angeles CA: University of California Press.

from apex predators to herbivores: Friedlander and DeMartini (2002); Knowlton and Jackson (2008); Sandin et al. (2008).

from vertebrates to invertebrates: Jackson and Sala (2001).

an inevitable problem of multiple ecosystem states: Scheffer et al. (2001); Knowlton (2004).

significantly degraded from their pristine state: Jackson (2008).

attempt to reverse degradation once it has occurred: Jackson et al. (2001); Scheffer et al. (2001).

managers were able to act quickly and on appropriately very large scales to ward off total collapse: Safina, C., A.A. Rosenberg, R.A. Myers, T.J. Quinn II, and J.S. Collie (2005), U.S. ocean fish recovery: Staying the course. *Science* 309: 707–708; Newman, M.J.H., G.A. Paredes, E. Sala, and J.B.C. Jackson (2006), Structure of Caribbean coral reef communities across a large gradient of fish biomass. Ecology Letters 9: 1216–1227; Rosenberg, Alexander and Cournane, this volume.

partial recovery of benthic ecosystems and fisheries in Tampa Bay and the Black Sea: Rabalais, N.N., R.E. Turner, and W.J. Wiseman Jr. (2002), Gulf of Mexico hypoxia, a.k.a. "The Dead Zone." *Annual Review of Ecology and Systematics* 33: 235–263.

due to the storage of vast quantities buried in sediments that recharge excess nutrients to the ecosystem: Meyer-Reil, L.A., and M. Koster (2000), Eutrophication of marine waters: Effects of benthic microbial communities. *Marine Pollution Bulletin* 41: 255–263.

Page 203

that can help sustain the commitment to stay the course in others: Pandolfi et al. (2005); Safina et al. (2005).

which take so much longer to grow and reproduce: Jackson (1991); Newman et al. (2006); Knowlton and Jackson (2008).

Epilogue

Page 206

which have devastated reef corals in the past: Sweatman, H. (2008), No-take reserves protect coral reefs from predatory starfish. *Current Biology* 18(14): R598–R599.

marine protected areas confer greater resistance to the effects of global climate change: Knowlton, N., and J.B.C. Jackson (2008), Shifting baselines, local

impacts, and global change on coral reefs. *PloS Biology* 6(2): 215–220, e54 (doi:10.1371/journal.pbio.0060054).

alternative strategies based on longstanding local traditions for management have proven effective in Melanesia and elsewhere: Foale, S. (2006), The intersection of scientific and indigenous ecological knowledge in coastal Melanesia: Implications for contemporary marine resource management. International Social Science Journal 58(187): 129–137.

CONTRIBUTORS

Karen E. Alexander is a historical fisheries scientist at the University of New Hampshire. She has coordinated the Gulf of Maine Cod Project since 2002, using her degrees in mathematics and history to facilitate communications between the project's scientists and historians. Co-editor of *Journal of a Cruise* by Captain David Porter, she also served as an advisor on a US Supreme Court case as well as on several documentary films, and currently writes on maritime history and fisheries science.

W. Jeffrey Bolster is an associate professor of history at the University of New Hampshire, best known for his prize-winning *Black Jacks: African American Seamen in the Age of Sail*. A maritime historian increasingly fascinated by changes in the sea, his research has shifted to marine environmental history. Bolster is part of the interdisciplinary Gulf of Maine Cod Project, and has contributed to their papers in *Frontiers in Ecology and the Environment* and *Fish and Fisheries*, among other journals. His solely authored papers have appeared in *Environmental History* and *The American Historical Review*. He is currently writing a book on the environmental history of the northwest Atlantic in the age of sail.

Francisco Chavez is a biological oceanographer with interests in how climate variability and change regulate ocean ecosystems on local and basic scales. He was born and raised in Peru, has a BS from Humboldt State and a PhD from Duke University. He was one of the first members of the Monterey Bay Aquarium Research Institute (MBARI) where he pioneered time series research and the development of new instruments and systems to make this type of research sustainable. Chavez has authored or co-authored more than a hundred peer reviewed papers, with ten in *Nature* and *Science*. He is a past member of the National Science Foundation Geosciences Advisory Committee, has been heavily involved in the development of the US Integrated Ocean Observing System (IOOS), and is a member of the governing board of the Central and Northern California Coastal Ocean

Observing System (CeNCOOS) and the Science Advisory Team for the California Ocean Protection Council. Chavez is a Fellow of the American Association for the Advancement of the Sciences, honored for distinguished research on the impact of climate variability on oceanic ecosystems and global carbon cycling. He was named Doctor *Honoris Causa* by the Universidad Pedro Ruiz Gallo in Peru in recognition of his distinguished scientific career and for contributing to elevate academic and cultural levels of university communities in particular and society in general.

Jamie M. Cournane, a postdoctoral research fellow at the University of New Hampshire and Environmental Defense Fund, currently serves on the Atlantic Herring Plan Development Team of the New England Fishery Management Council. She has worked most recently on mapping hot spots of river herring bycatch by trawlers and seiners that target Atlantic herring. For her doctoral work, she assessed spatial patterns of groundfish biodiversity in the Gulf of Maine and Georges Bank over the past hundred years and emphasized that historical perspectives provide baselines to measure success in the current spatial management of fisheries.

Jon M. Erlandson is an archaeologist, professor of anthropology, and Knight Professor of Arts and Sciences at the University of Oregon (UO), where he directs the Museum of Natural and Cultural History. He earned his PhD from the University of California, Santa Barbara in 1988, and taught at the University of Alaska–Fairbanks before joining the UO faculty in 1990. With sixteen books and more than two hundred scholarly articles published, Erlandson's research focuses on the origins and development of maritime societies, human migrations, the peopling of the Americas, and the historical ecology of marine fisheries and coastal ecosystems.

David B. Field is currently an assistant professor in marine sciences at Hawaii Pacific University. He has worked extensively in reconstructing past changes in climate and marine populations from laminated marine sediments as well as calibrating and determining the fidelity of the fossil records with historical records and plankton tows. He has used planktonic foraminifera from the Santa Barbara Basin to distinguish the warming trend from natural variability in the California Current. Investigations with fish debris have been used to infer past variations in fish populations. Field has also lived and worked in Peru as part of an international group of investigators (Paleopeces) reconstructing climate and ecosystem change in the Peru-Chile Current.

Marah J. Hardt, founder of OceanInk, is a research scientist, writer, and consultant. A coral reef ecologist by training, she keeps one foot wet in the field, while the other roams the worlds of creative storytelling and problem-solving, with a focus on ocean conservation and climate change issues. Her interdisciplinary background and effective communication skills allow her to work with diverse thought leaders, from scientists to social entrepreneurs, to create innovative solutions to pressing conservation problems. Her articles have appeared in academic and popular media, such as *The American Prospect, Ecology Letters*, and *Scientific American*.

Jeremy B. C. Jackson is the Ritter Professor of Oceanography and Director of the Center for Marine Biodiversity and Conservation at the Scripps Institution of Oceanography in La Jolla, California, and Senior Scientist Emeritus at the Smithsonian Tropical Research Institute in the Republic of Panama. Previously he was Professor of Ecology at Johns Hopkins University. He is the author of more than 150 scientific publications and author or editor of seven books. His research includes human impacts on the oceans and the ecology and paleoecology of tropical and subtropical marine ecosystems. Dr. Jackson is a Fellow of the American Academy of Arts and Sciences and the American Association for the Advancement of Science and recipient of numerous international prizes and awards. His work on overfishing was chosen by *Discover* magazine as the outstanding scientific achievement of 2001.

Carina B. Lange was born and raised in Buenos Aires, Argentina. She received her undergraduate degree in biology from the University of Buenos Aires, followed by a doctorate in marine biology at the same university. She left Argentina in 1984 with a UNESCO scholarship for the Scripps Institution of Oceanography in California. There, her main research focused on laminated Quaternary sedimentary records as well as phytoplankton time-series studies from off California. Lange is the author of numerous scientific articles on diatom ecology and taxonomy, diatom fluxes to the seafloor and preservation in the sediments, as well as paleoreconstructions from sedimentary archives worldwide. Since September 2001 Lange has been a professor of oceanography at the University of Concepción, Chile, where she is involved in graduate teaching, scientific research, and academic administration. She is also the director of the Center for Oceanographic Research in the eastern South Pacific (COPAS), hosted at the University of Concepción, and the leader of the project Oceanographic Applications for the Sustainable Economic Development of the Southern Region of Chile. In addition, Lange holds a position as research associate at Scripps. Currently, one of her main lines of research focuses on paleoceanographic and paleoclimate changes of the Late Quaternary and the Holocene in the eastern South Pacific.

William B. Leavenworth is a marine environmental historian for the Gulf of Maine Cod Project at the University of New Hampshire, and a professional sailor. In 1976 he shipped on the tall ship *Gazela Primeiro*, and stayed in the sailing profession for twelve years. Later he obtained a doctorate in history from the University of New Hampshire, with a dissertation in colonial New England maritime environmental history. His firsthand knowledge of the sea has been key to interpreting historical fisheries records, and he has been with the Gulf of Maine Cod Project since 2001. In addition to scholarly articles concerning historic cod fisheries in the Northwest Atlantic, he has published short stories and poetry.

Heike K. Lotze is an assistant professor in marine biology at Dalhousie University in Halifax, Canada, and holds the Canada Research Chair in Marine Renewable Resources. Trained in marine ecology and biological oceanography, she has a strong interest in how human activities alter marine populations and ocean ecosystems. In her research, she tries to reconstruct the long-term history of human-induced changes in the ocean and analyze the consequences on the structure and functioning of marine ecosystems and the services they provide for human well-being. Lotze received a masters in biology in 1994 and a doctorate in biological oceanography in 1998 from Kiel University in Germany. She has worked as a postdoctoral fellow and research associate at Dalhousie University in Halifax, Canada, and the Alfred-Wegner Institute for Polar and Marine Research in Bremerhaven, Germany, and has participated in several working groups at the National Center for Ecological Analysis and Synthesis in Santa Barbara, California.

Alec D. MacCall has worked on fish population dynamics for more than forty years. He has focused on assessment and management of California's coastal pelagics and long-lived groundfish, and has contributed to the successful rebuilding of several depleted stocks (California sardines, Pacific mackerel, and bocaccio rockfish). He has written extensively, and has contributed to assessment methodologies, ecological modeling, effects of interdecadal climate variability, and development of adaptive harvest control policies. His current work focuses on developing data-limited stock assessment methodologies. He received a doctorate in oceanography from the University of California's Scripps Institution of Oceanography in 1983. He joined the National Marine Fisheries Service's Southwest Fisheries Science Center (NMFS-SWFSC) in 1982, after working twelve years for the California Department of Fish and Game. From 1988 to 1997 he was director of the NMFS-SWFSC Tiburon Laboratory. He currently holds the position of Senior Scientist at the NMFS-SWFSC Laboratory in Santa Cruz, California.

Loren McClenachan is a NSF International Postdoctoral Fellow at Simon Fraser University. She received her doctorate in marine biology from the Scripps Institution of Oceanography, where she researched historical changes in tropical marine ecosystems. Her current work addresses issues in historical ecology and marine conservation.

Richard D. Norris works on the evolutionary dynamics of ocean plankton using the marine microfossil record preserved in deep ocean sediments. These studies focus both on mechanisms of extinction and speciation as well as on biological responses to past periods of "extreme climate" and mass extinctions. In addition, he studies historical records of human impacts on modern marine ecosystems. Norris was an undergraduate in Earth Sciences at UC Santa Cruz, obtained a master of science at University of Arizona, Tucson, and a doctorate in geology at Harvard University. In between these academic programs he worked on the Condor Recovery Project for the State of California and served as director of the NRS Granite Mountain Reserve for the University of California. Following graduate training, he was a research scientist at Woods Hole Oceanographic Institution, on Cape Cod, Massachusetts, until he moved in 2002 to UC San Diego as a full professor at Scripps Institution of Oceanography.

Randy Olson earned his PhD at Harvard University and became a professor of marine biology before moving to Hollywood for his second career as a filmmaker. Since obtaining an MFA from the University of Southern California School of Cinema, he has written and directed the critically acclaimed films *Flock of Dodos: The Evolution-Intelligent Design Circus* (Tribeca 2006, Showtime) and *Sizzle: A Global Warming Comedy* (Outfest 2008), and co-founded the Shifting Baselines Ocean Media Project, a partnership between scientists and Hollywood to communicate the crisis facing the ocean. He is the author of *Don't Be Such a Scientist: Talking Substance in an Age of Style*.

Daniel Pauly is a French citizen who completed his high school and university studies in Germany; his doctorate (1979) is in fisheries biology, from the University of Kiel. After many years at the International Center for Living Aquatic Resources Management (ICLARM), in Manila, Philippines, Pauly became in 1994 a professor at the Fisheries Centre of the University of British Columbia, of which he was the director from 2003 to 2008. Since 1999, he has also been Principal Investigator of the Sea Around Us Project, funded by the Pew Charitable Trusts, and devoted to studying and documenting the impact of fisheries on the world's marine ecosystems (www.seaaroundus.org). The concepts, methods, and software he codeveloped, documented in more than five hundred publications, are used throughout the world, following multiple courses and workshops given in four languages on all five continents. This applies especially to the Ecopath modeling approach and software (www.ecopath.org) and FishBase, the online encyclopedia of fishes (www.fishbase.org). This work is recognized in various profiles, notably *Science* (April 2002), *Nature* (January 2003), *New York Times* (January 2003), and by numerous awards, notably the International Cosmos Prize, Japan (2005), the Volvo Environmental Prize, Sweden (2006), and the Excellence in Ecology Prize, Germany (2007).

Stephen R. Palumbi is director of the Hopkins Marine Station and the Jane and Marshall Steele Jr. Professor of Marine Science at Stanford University. He has lectured extensively on human-induced evolutionary change, has used genetic detective work to identify whales for sale in retail markets, and is working on new methods to help design marine parks for conservation. His latest book is an unusual environmental success story called *The Death and Life of Monterey Bay: A Story of Revival*. Palumbi holds a doctorate from the University of Washington and a B.A. from Johns Hopkins University. He has received numerous awards for research and conservation, including a Pew Fellowship in Marine Conservation. He lives in Pacific Grove, California, and is based at Stanford's Hopkins Marine Station.

Andrew A. Rosenberg is Senior Vice President for Science and Knowledge at Conservation International and a professor in the Institute for the Study of Earth, Oceans, and Space at the University of New Hampshire where, prior to April 2004, he was dean of the College of Life Sciences and Agriculture. From 2001 to 2004, he was a member of the US Commission on Ocean Policy and continues to work with the US Joint Ocean Commissions Initiative. Rosenberg was the deputy director of NOAA's National Marine Fisheries Service from 1998 to 2000, the senior career position in the agency, and prior to that he was the NMFS Northeast Regional Administrator. Rosenberg's scientific work is in the field of population dynamics, resource assessment, and resource management policy. He holds a bachelor of science in fisheries biology from the University of Massachusetts, an master of science in oceanography from Oregon State University, and a doctorate in biology from Dalhousie University.

Kaustuv Roy is a professor at the University of California, San Diego. His research focuses on macroecology, macroevolution, and conservation biology.

Carl Safina writes about how the ocean is changing. A MacArthur fellow, Pew fellow, and Guggenheim fellow, he is adjunct professor at Stony Brook University and president of Blue Ocean Institute. His books include *Song for the Blue Ocean* and *The View from Lazy Point: A Natural Year in an Unnatural World*.

Enric Sala obtained his doctorate from the University of Aix-Marseille, France. After an early career at the Scripps Institution of Oceanography in San Diego, California, he returned to Spain to sit on the Spanish National Council for Scientific Research (CSIC). Now based in Washington, DC, as the head of National Geographic's global marine conservation initiative, Sala dives all over the world to explore marine ecosystems and promote their conservation.

Paul E. Smith, now retired, served as Supervisory Fisheries Biologist at NOAA and was adjunct professor of biological oceanography at the Scripps Institution of Oceanography.

Tim D. Smith is a fishery biologist turned environmental historian, interested in the effects of harvesting on long-lived species, including fish and seals, and especially cetaceans. He is retired from NOAA and has been active in the History of Marine Animal Populations project, part of the Census of Marine Life.

U. Rashid Sumaila is director of the Fisheries Centre at the University of British Columbia. He specializes in bioeconomics, marine ecosystem valuation, and the analysis of global issues such as fisheries subsidies; illegal, unreported, and unregulated fishing; and the economics of high and deep seas fisheries. He has published articles in several journals, including *Science, Nature*, and the *Environmental Economics and Management*. Sumaila has received invitations to give talks at the United Nations, the White House, the Canadian parliament, and the British House of Lords. His work has been cited by, among others, the *Economist*, *Boston Globe, International Herald Tribune, Financial Times*, and *Globe and Mail*.

Daniel Vickers works in the fields of Early American, Atlantic, and maritime history. He taught at the Memorial University of Newfoundland for fifteen years before moving to the University of California, San Diego in 1999 and to the University of British Columbia in 2006, where he served as head of the department until 2011. His first book,*Farmers and Fishermen: Two Centuries of Work in Essex County, Massachusetts, 1630–1850*, won the Dunning Prize of the American Historical Association, and he has recently published a second, *Young Men and the Sea: Yankee Seafarers in the Age of Sail*. He is also the editor of the *Blackwell Companion to Colonial American History* and *The Autobiography of Ashley Bowen*.

Christine R. Whitcraft is an assistant professor in the Biological Sciences Department at California State University, Long Beach (CSULB). She earned a bachelor of art at Williams College, a doctorate at Scripps Institution of Oceanography, and was a CALFED Bay-Delta Program postdoctoral fellow with the San Francisco Bay National Estuarine Research Reserve before coming to CSULB in 2008. As a biological oceanographer and wetlands ecologist, she teaches a variety of ecology classes and researches the impacts of human activities on coastal ecosystems.

INDEX

Note: page numbers followed by 'f' or 't' refer to figures or tables, respectively.

Island Press | Board of Directors